City Worlds

edited by
Doreen Massey, John Allen and Steve Pile

London and New York

in association with

The Open
University

First published 1999 by Routledge; written and produced by The Open University

11 New Fetter Lane, London EC4P 4EE

Simultaneously published in the USA and Canada

by Routledge

29 West 35th Street, New York, NY 10001

© The Open University 1999

The opinions expressed are not necessarily those of the Course Team or of The Open University.

Edited, designed and typeset by The Open University

Index compiled by Isobel McLean

Printed in Great Britain by T.J. International, Padstow, Cornwall

British Library Cataloguing in Publication Data

A catalogue record for this book is available from The British Library

Library of Congress Cataloging in Publication Data

A catalogue record for this book has been requested

ISBN 0-415-20069-5 (hbk)

ISBN 0-415-20070-9 (pbk)

1.1

CONTENTS

THE OPEN UNIVERSITY COURSE TEAM

John Allen *Senior Lecturer in Economic Geography*

Sally Baker *Education and Social Sciences Librarian*

Melanie Bayley *Editor*

Andrew Blowers *Professor of Social Sciences (Planning)*

Christopher Brook *Lecturer in Geography*

Deborah Bywater *Project Controller*

David Calderwood *Project Controller*

Margaret Charters *Course Secretary*

Allan Cochrane *Professor of Public Policy*

Lene Connolly *Print Buying Controller*

Michael Dawson *Course Manager*

Margaret Dickens *Print Buying Co-ordinator*

Nigel Draper *Editor*

Janis Gilbert *Graphic Artist*

Celia Hart *Picture Research Assistant*

Caitlin Harvey *Course Manager*

Steven Hinchliffe *Lecturer in Geography*

Teresa Kennard *Co-publishing Advisor*

Siân Lewis *Graphic Designer*

Michèle Marsh *Secretary*

Doreen Massey *Professor of Geography*

Eugene McLaughlin *Senior Lecturer in Criminology and Social Policy*

Gerry Mooney *Staff Tutor in Social Policy*

Eleanor Morris *Series Producer, BBC/OUPC*

John Muncie *Senior Lecturer in Criminology and Social Policy*

Ray Munns *Cartographer*

Steven Norris *Graphic Designer (Cover)*

Kathy Pain *Staff Tutor in Geography*

Steve Pile *Lecturer in Geography and Course Team Chair*

Michael Pryke *Lecturer in Geography*

Jenny Robinson *Lecturer in Geography*

Kathy Wilson *Production Assistant, BBC/OUPC*

External Assessor

John Solomos *Professor of Sociology, University of Southampton*

External Contributors

Ash Amin *Author, Professor of Geography, University of Durham*

Stephen Graham *Author, Reader in the Centre for Urban Technology, University of Newcastle upon Tyne*

Kerry Hamilton *Author, Professor of Transport, University of East London*

Mark Hart *Tutor Panel, Reader in Industrial and Regional Policy, University of Ulster*

Susan Hoyle *Author, Research Associate in the Transport Studies Unit, University of East London*

Linda McDowell *Author, Director for the Graduate School of Geography and Fellow at Newnham College, University of Cambridge*

Ian Munt *Tutor Panel, Researcher, London Rivers Association*

Phil Pinch *Tutor Panel, Senior Lecturer, Geography and Housing Division, South Bank University*

Jenny Seavers *Tutor Panel, Research Fellow, Centre for Housing Policy, University of York*

Nigel Thrift *Author, Professor of Geography, University of Bristol*

Sophie Watson *Author, Professor of Urban Cultures, University of East London*

Preface

City Worlds is one of a series of books, entitled *Understanding Cities*, that takes a
new look at cities. The standard approach to thinking about the future of cities is
to consider them as free-standing and geographically discrete places that rise and
fall as a result of their strategic location, their economic viability or their political
power. Typically, the history of cities is charted *from* the rise and fall of ancient
cities (such as Athens and Rome), *through* the rise of mediaeval cities (such as
Antwerp and Naples), *onto* the spectacular growth of cities during the industrial
revolution and the age of empire (such as London and Paris), and finally *to* the
sprawling 'post-modern' cities of today (normally exemplified by Los Angeles).
The future of cities is then extrapolated on the basis of this historical vision –
usually that urban areas will continue to sprawl across the surface of the Earth,
eventually joining up to form '100-mile' or 'mega-'cities. The analysis developed
in this series is, however, radically different.

In order to understand cities, we argue that it is necessary to rethink their
geography. This is more than simply extending the range of cities considered
beyond the London–New York–Los Angeles axis, whether from São Paulo to
Sydney, from Manchester to Moscow. It also involves using a geographical
imagination to understand how cities are produced, on the one hand, in a
context of social relations that stretch *beyond* the city and, on the other, by the
intersection of social relations *within* the city. This argument has widespread
implications for our understanding of cities. These implications are
comprehensively investigated in the three books that comprise this series, *City
Worlds, Unsettling Cities* and *Unruly Cities?*. Through the series, we tease out, for
example, the ways in which cities bring people from different backgrounds into
close proximity; how the juxtaposition of different people and activities in cities
can change and alter social interactions; how these juxtapositions can result in,
or result from, urban conflicts and tensions; how different parts of cities are
connected to, or disconnected from, other cities; how people network within and
between cities.

Developing these arguments shows that the city cannot be thought of as having
one geography and one history (and therefore one future). Instead, cities are
characterized by their openness: to new possibilities, and to new interactions
between people. This book series gradually reveals the difficulties and paradoxes
that the unavoidable openness of cities presents, as different histories and
geographies intersect and overlap. Significantly, then, it is these issues that must
be understood, if people are to learn to live in our increasingly urbanized world.

A few words about the books themselves. Each book is self-contained: each
chapter provides all relevant supporting readings and materials to enable the
reader to come to grips with a new understanding of cities. The three textbooks
are, also, a significant component of the Open University third level course,

DD304 *Understanding Cities.* The book series is complemented by a television series, *City Stories,* and other materials, including audio-cassettes and course guides.

The television series picks up the themes of the three books through case-study material from across the globe – visiting Sydney, Singapore, Kuala Lumpur, Moscow and Mexico City. In particular, the television series develops a line of argument about the different ways in which groups within cities locate themselves within specific networks of social relationships. Local contributors comment on and analyse their own cities, highlighting the issues of importance for their urban futures. The television series is integral to the Open University course and provides a wealth of examples to illustrate the themes and the experiences of living in cities around the world. (Details of how to obtain copies of the TV programmes are available from Open University Worldwide Ltd.: see p.ii.)

Open University courses are produced through extensive and intensive discussions amongst academic authors, a panel of experienced tutors, an external assessor, editors, designers, BBC producers, a course manager and, not least, secretarial staff (listed on p.vi). Every component of the course has been subjected to wide-ranging discussion and critical debate. At every stage, this has led to improvements in the materials, which have benefited from the academic and distance-teaching expertise accumulated at The Open University.

The end-product of this process is to produce textbooks that have a specific style. In particular, authors have sought to make these texts *interactive.* Readers are asked to become involved in the problem of understanding cities, rather than simply to digest the material. Readers are provided, too, with a wide range of readings and extracts, which they can use to enhance their own appreciation of the problems that cities confront. Certain ideas become important to this understanding of cities – and these are registered in two basic ways. First, key points are emphasized using bullet points; and, second, core ideas, that are discussed in detail, are indexed in **bold**. Meanwhile, each book's last chapter looks back over the book as a whole and draws out the key themes that have been discussed. For readers who wish to use the series as a whole to further their understanding of cities, there are a number of references backwards and forwards to chapters in other books in the series; these are easily identifiable because they are printed in **bold** type. This is particularly important since chapters are intended to build on previous discussions. By integrating the material in this way, it is hoped that readers will be able to pull out the themes that run across the series for themselves. Through this interactive and integrated approach, it is hoped that readers will develop their own understanding of cities and ultimately be able to transfer that understanding to other cities, even to other situations.

It only remains to acknowledge those who have worked hard and with such enthusiasm to produce this book series. Most obviously, these texts – and the course of which they are a part – would not be possible without the academic consultants and the tutor panel. All have helped to shape the chapters of others, while also responding constructively to suggestions and advice. The external assessor, John Solomos, provided invaluable intellectual guidance, insights, support and encouragement at every stage of the development of the books and the related course materials. The tutor panel, comprising Mark Hart, Ian Munt and Jenny Seavers, have been rigorous in their comments on the drafts as the chapters developed. Finally, Phil Pinch, who wrote the Course Guide, generously advised on the development of the course as a whole.

Producing an excellent textbook series does not end with the writing. The production process has been co-ordinated and kept on track by Deborah Bywater and David Calderwood. We have been shrewdly supported by our editors, Melanie Bayley and Nigel Draper, who scrupulously scrutinised the chapters before publication. Meanwhile, the design of the books and chapters has been thoughtfully overseen by Siân Lewis. Our cartographer and graphic artist, Ray Munns and Janis Gilbert, developed and drew the maps and diagrams. Together, they have turned the drafts into the excellent books they undoubtedly are. Finding illustrations for this course has proved, at times, a frustrating task, but Sally Baker and Celia Hart have been good humoured throughout. The production of textbooks at the Open University means a never-ending stream of requests for word-processing and the distribution of drafts, often at very short notice. With good grace and efficiency, the course team has been supported by Michèle Marsh and, in particular, the course secretary, Margaret Charters. Within the Faculty of Social Sciences, Peggotty Graham has helped to ensure that the books were produced to high standards, while – at Routledge – Sarah Lloyd has been a constant source of good advice, flexibility and encouragement.

At the centre of this process are the course managers who have worked on this series of books. We have been lucky enough to have gained from the expertise of Christina Janoszka and Mike Dawson. However, it is Caitlin Harvey who has seen the books through the most important stages. Caitlin has never failed to keep her head, even while all about her were losing theirs. Keeping everything together has meant that course team meetings have been intellectually lively and thoroughly thought-provoking. We hope that the vitality and diversity of these debates shine through in these textbooks, which offer a new way of *Understanding Cities*.

Steve Pile
on behalf of The Open University Course Team

Introduction

City Worlds presents both an overview and the nub of the argument in the *Understanding Cities* series. It is our contention that, at the turn of this millennium, the problems and possibilities which cities embody and express are among the most important issues facing the planet. For the first time in history more than one half of humanity will be living not just in cities but in mega-cities. Every day, tens of thousands of new people arrive in cities around the world. Never before have we experienced such a geography, such a concentration of the world's population. What possibilities and problems does this new geography bring? And how might we go about analysing and addressing them?

Twice now already, we have referred to 'possibilities' and 'problems' together. And that characteristic of double-sidedness is one of the things we do know about cities. On the 'possibility' side, there is the city as the apex of civilization, the birthplace of citizenship, the City on the Hill. On the 'problem' side, there is the city of poverty, mayhem and threat; the bursting of the bounds of social control. On the one hand, cities are the crucibles of the new, places of mixing and the creation of new identities; they are the cradles of new ideas. On the other hand, that very process of the coming together of different peoples can create conflict, intolerance and violence. These contrasting images of the city overlap and play off one another, perhaps because the city comprises so many worlds. Taking these contrasts together, it would appear that the city is nothing if not ambiguous. This ambiguity – the coexistence of, and tension between, possibilities and problems – runs throughout both this book and the series as a whole. Indeed, it is perhaps this very dynamic ambiguity which is behind the fact that, so often throughout history, cities have been at the forefront of social change. The question, of course, is what kind of social change. Because of the increasing size of cities, and the numbers of people involved, the question has never been more important.

Cities, then, are of enormous significance. But it is also true that there has been a lot written about them. Our approach in this book, and in this series, is both to take account of and consider a range of that existing material and also to present a distinctive angle. What we want to do is to explore and exemplify the particular contribution which can be made to the study of cities by adopting a specifically spatial imagination. *City Worlds* establishes the foundations of this approach. In part, this means that we are conscious of the global reach of the stories we tell about cities. In part, we seek to uncover the ways in which the tensions of the city that are produced by social relations are spatially constituted, both within and beyond the city. In part it means that the range of 'issues' we draw in is very wide: from life on the streets to the economics of neo-liberalism, from issues of 'community' to questions of environmentalism. But more than this, it means that we propose a particular, and particularly spatial, way of imagining cities.

By tracing the cities' untold spatial relations, *City Worlds* is able to tell new stories about cities. In these stories, a new understanding of cities emerges. There are two aspects to this which are of particular importance. First, cities are places of particularly intense social interactions, places of a myriad social juxtapositions. They are, as we have said, possibly the prime sites of social mixing, of the coming together and interacting of distinct narratives. And out of this mixing and interaction new stories, new narratives, are born. Second, we argue, it is impossible to tell the story of any individual city without understanding its connections to elsewhere. Cities are essentially open; they are meeting places, the focus of the geography of social relations. They pull into themselves, are the foci for, the world's networks of power and information, the complexities of the movements of finance and trade, the social disruptions and the personal journeys of migrants, the centres of power of media, culture and communications, and the focus of the imaginations of millions. It is, we argue, this setting of cities within a wider geometry of social relations which helps to explain their dynamism and their complexity.

It is, moreover, this ever-shifting combination of openness and heterogeneity which lies behind the specificity both of what cities have to offer and of the challenges which they pose. In this first volume of the series we therefore also begin to highlight and explore some of the essential tensions of urbanism: between community and difference, between movement and settlement, between order and disorder. And by starting to excavate these issues we can also begin to raise questions about the future of cities, problems and possibilities which revolve around issues of inequality, tolerance, democracy and sustainability.

Doreen Massey, John Allen and Steve Pile

CHAPTER 1
What is a city?

by Steve Pile

1 *Introduction*

1.1 IMAGES OF THE CITY

It is easy to ask the question 'what is a city?', but – as you might have already anticipated – less easy to answer it. Let us think about images of the city to see why this might be so. Tour operators regularly advertise holidays to cities, where you can revel in the entertainments and sights that they have to offer. Meanwhile, other holiday advertisements proclaim the benefits of leaving the city and getting away from it all. It is not just holidaymakers and the tourist industry that have some sense of what cities are like. Commonly, people will talk of going to the city to get this or that; or they will say that they come from the city, whether because it was their birthplace or where they live. Airlines seek to move high-flying executives between cities: to make contacts, to build up business. Others, meanwhile, are down and out in the city. Moreover, governments tend to locate the major institutions of state in capital cities – some countries even build new cities for the purpose.

In people's imaginations, there seems to be a clear idea of what the city is – its opportunities, its benefits and its dangers. However, these images of the city are very different: holidaymakers, business managers, city-dwellers, government officials, all have subtly different images of what the city is like, what the city offers them and what they want to do in cities. So, people might have very definite answers when asked what a city is: it is there for pleasure and fun, for cultural events, for business and profit, for home and work, for administration and government, and so on. But, if we want to understand what the city is, then we will have to think about what we learn when these images are put side by side.

ACTIVITY 1.1 Before reading on, reflect for a moment on the images that *you* have of a city. Make a list of the kinds of images that are springing to your mind. Next, make a separate list of the kinds of things that can *only* be found in cities. ◆

My list of urban features began with physical features such as houses, housing estates, streets, shops, hotels, hospitals, museums, traffic, libraries, cathedrals, soup-kitchens, restaurants, and so on – and so on, so I stopped writing after a while. On the other hand, the number of things that can only be found in cities appears to be much smaller: I thought of skyscrapers, underground railways, street lighting (maybe), and not much else. We can see quite quickly that many features of the city can be found outside the city. There are housing estates and hospitals in rural areas, as well as shops and museums, but these tend to be smaller in scale than their equivalents in the city. Maybe what is distinctive about cities is, in essence, a question of size. Is what a city is determined by the scale, say, of its office-blocks or of its housing estates? In part. But a small village with a huge museum or hospital or even a large housing estate would not be a city. Cities have something more than simply 'largeness'. Sure, cities are big, but their size is related to the way in which they *combine* most – if not all – of the features on our lists.

If cities are a combination of many features, then this leads to another question: does the city have to have a particular blend of elements to be a city? Yes and no. Very few cities have exactly the same mixture of things. Let us take one example: the skyscraper. A city does not have to have skyscraper to be a city, but skyscrapers are only found in cities. So, the skyscraper is a city feature, but not a feature of all cities. The list of things we find in cities does not appear to add up to what a city is. Maybe there is something more in this question about whether cities combine elements in distinctive ways.

Some time ago, the urban geographer Brian Robson asked himself a similar question about what the *urban environment* is like. His answer was this:

> The urban environment is a deceptively simple term. It conjures up images of crowded Oxford Street (in London) thronged with shoppers, the haunting engravings of Gustave Doré with their pictures of slum life under the railways arches of Victorian London, the pyramid skyscrapers on the skyline of New York, the endless similar, and similarly evocative, images. What it is that we are evoking in such images however, is very difficult to specify with any precision. We tend to conflate the physical and the human aspects together and, while we can say with some confidence what we mean by 'urban' in physical terms, it is much more difficult to spell out its social significance.

(Robson, 1975, p.184)

In Robson's view, the city is more than a collection (or combination) of images, but something that has social significance. This suggests that we need to think about the social significance of the items on our list of city features. Certainly, in Robson's images of the city, there are less tangible aspects such as (over)crowding, shopping and slum life, involving destitution, sickness and hopelessness. You might have noted down similar features on your list of images of the city. For many, it is the intangible qualities of *city life* that makes it distinctive, such as luxury and poverty, amenity and pollution, tradition and innovation, drudgery and novelty, order and disorder, thrills and spills, volatility and conflict, difference and indifference, public services and welfare provision, individual freedom and dependency on others. Clearly, some of these traits seem to be at odds with others: how can the city contain both order and conflict, freedom and dependency? Perhaps what cities are about is the attempt to deal with (or make the best of) these tensions.

Even if this is right, experiences of the city will differ from person to person as well as from group to group. Indeed, they will vary depending on the part of the city in which the person or group is located. Of course, people's feelings are not confined to cities: farm work can be every bit as monotonous and routine as factory work, country pursuits just as thrilling as the city's excitements; loneliness and neighbourliness, wealth and poverty are possible in cities and villages alike. However, the city tends to exaggerate such relationships, if only by bringing them into close proximity. And perhaps this makes matters worse (or better). Once more, what is distinctive about urban experiences might be a question of degree. From this perspective, city life is distinctive because its scale is larger and activities more intense than anywhere else. When we look back over these images, we find that we have raised similar questions about what a city is. The city contains many features and experiences, but simply cataloguing these has not proved enough to define the specific qualities of the city. That is, there is something distinctive about

● the scale and intensity of urban life, and

● the combination of urban elements.

However, we have also found that it is also necessary to consider

● the social significance of the city.

Perhaps we can begin to specify what the city is by taking somewhere we can agree to be a city? Let us consider this image of the skyline of Manhattan Island in New York, USA.

FIGURE 1.1 *Skyline of Manhattan, showing the twin towers of the World Trade Center*

This city skyline is one of the most famous in the world. Perhaps because few Hollywood film-makers can resist using it as a backdrop to the action: as the heroine clings by her fingernails to the upper reaches of the Statue of Liberty; as the hero chases the bad guys in a speed-boat towards the towering metropolis; as King Kong meets his doom, atop the Empire State Building. There is something about the skyline which itself evokes action, which evokes something of the action of cities themselves. Urban theorists, too, have also been awed by the sight of Manhattan's sky-scraping buildings. For example, Kevin Lynch describes Manhattan like this:

> The image of the Manhattan skyline may stand for vitality, power, decadence, mystery, congestion, greatness, or what you will, but in each case that sharp picture crystallizes and reinforces the meaning.

> (Lynch, 1960, pp.8–9)

About a quarter of a century later the French social theorist, Michel de Certeau, would describe the scene this way:

> Seeing Manhattan from the 110th floor of the World Trade Center. Beneath the haze stirred up by the winds, the urban island, a sea in the middle of the sea, lifts up the skyscrapers over Wall Street, sinks down in Greenwich, then rises again to the crests of Midtown, quietly passes over Central Park and finally undulates off into the distance beyond Harlem. A wave of verticals. Its agitation is momentarily arrested by vision. The gigantic mass is immobilized before the eyes. It is transformed into a texturology in which extremes coincide – extremes of ambition and degradation, brutal oppositions of races and styles, contrasts between yesterday's buildings, already transformed into trash cans, and today's urban irruptions that block out its space.

> (de Certeau, 1984, p.91)

In each of these statements, the writers are trying to get at the way in which the skyline of Manhattan conveys something of the city's social significance – but with every word they add, this becomes more elusive. Is New York vital, powerful, decadent, mysterious, congested, great; is it ambitious and degraded, a contrast of ethnicities, styles, yesterdays and todays? It is surely all of these things, and more. For de Certeau, the skyscrapers of New York seem to evoke for him the image of a stormy sea, as the height of the buildings rise and fall, as if driven by a tremendous storm. In these waves, he can see some buildings rising and others falling away. Each building representing a yesterday that enabled it to be built and a today that might tear it to the ground – to replace it with something better, something taller. Behind the buildings are the storms of social processes that build; that tear down; that build again.

In both these descriptions of Manhattan, the authors are also trying to capture something of the intensity of the city. For de Certeau at least, New York is born on the stormy seas of money; the city is built on the circulation and use of capital. Indeed, skyscrapers are very expensive, both to tear down and to build. Invariably, investors have sited their skyscrapers in cities – and, as a result, the city's image is associated with the vitality, prestige and dynamism of such large scale spending. There might be many reasons for building skyscrapers and for siting them in specific cities (and not others), but ultimately de Certeau is asserting that the city is built on – and is stained by – money (even if it is borrowed!).

It matters, then, how money circulates between and within cities, where money accumulates, where people decide to invest it and how they choose to spend it. Surprisingly, then, today some of the tallest buildings are not in the capital cities of the richer countries of the world – such as New York or London or Paris – but in the seemingly poorer

FIGURE 1.2
Petronas Towers, Kuala Lumpur

countries, in cities such as Kuala Lumpur (see Figure 1.2). And this is important precisely because it changes the image of KL (as Kuala Lumpur is commonly known): so, by building the Petronas Towers, the city intends to rival New York. But does it?

For de Certeau, the city is much more than the outcome of decisions to spend enormous sums putting up very tall buildings. He refuses to be entirely captivated by the seas of buildings that confront him. Instead he begins to look down from the top of the World Trade Center to the streets, to try to pick out what it is that people are actually doing on the ground. But de Certeau cannot make them out. The city seems immobilized before his eyes. On the streets, however, we can imagine the hustle and bustle; the noise of the yellow cabs hooting and the pollution billowing from their exhausts; the crowds of people waiting patiently at street corners to cross the roads and the people thick on the platforms of the underground. On the streets, people are everywhere: walking, jostling, standing, looking, shouting, begging, shopping, and … doing whatever they're doing. We can imagine millions of people, all with their own stories to tell; the city contains a world of possibilities (see Figure 1.3).

FIGURE 1.3   *Advertisement for a holiday in New Zealand showing a street scene of Oxford Street, London*

9

However, the streets de Certeau cannot quite make out from the rooftop of Manhattan are of a particular kind. The streets of New York might be very different from those elsewhere (see Çelik, Favro and Ingersoll, 1994). Indeed, the images we implicitly draw on when we use the word 'city' can stem from a very limited repertoire of cities. Thinking about street life in other cities can tell us more about the kinds of things that go on, partly by showing that the city brings together people, commodities, beliefs, money, and so on, from many different places.

ACTIVITY 1.2 Compare the images of the streets in Figure 1.4. What strikes you as being distinctive about each city? What features crop up in different places? ◆

(a)

(b)

FIGURE 1.4 *Chinatowns*

(c)

FIGURE 1.4
Street markets

(d)

In Figure 1.4 (a) and (b) we can see images of Hong Kong and of Chinatown in San Francisco. It is just about possible to tell which is which by looking at details, such as the architectural styles of the buildings and other shop signs. Otherwise, both these streets are dominated by Chinese symbols and customs. Meanwhile, in Figure 1.4 (c) and (d), we can see differing constructions of the market stalls, in this case in New Delhi and in Tower Hamlets, London. Taken together, we can imagine many different kinds of people participating in the *street life* of these cities: from the casual tourist wandering into a restaurant, and the rich who purchase expensive goods in élite stores, to the idle browser picking up the odd bargain from a street market, or the shopper who buys their weekly food supply at a local specialist foodstore, and the shoplifter. And we can multiply these stories to imagine as many streets and as many reasons for moving through those streets as we can. And we can add more buildings, more kinds of transport (both on the roads, and also on tracks, above and below the ground), more kinds of goods and services (made concrete in constructions such as sewage plants, churches, hotels, hospitals, police-stations, fire-stations, houses, and the like), more kinds of experiences (from witnessing spectacular events to watching television alone, from chatting with friends to arguing with the boss). More stories can be told, it might be said, until the cows come home. As we listen to the clamour of all these stories, what emerges is that what is distinctively urban is that there are *so many* stories.

Up to now, we have shown that cities contain

- many different physical features, and
- many different experiences, about which many stories can be told.

What is becoming clear is that it is important how these features and experiences are combined within cities. In order to understand this, it is important to ask how different aspects of the city are produced in the city; and, how these different aspects of the city are brought together (or kept apart). To take forward the question 'what is a city?', it is necessary to begin to address these questions.

1.2 SO, WHAT IS A CITY?

In asking the apparently simple question 'what is a city?', we ended up with a seemingly endless list of physical features, human experiences and urban images. However, from what we have seen so far, what is at issue is

- how seemingly essential or specific elements of the city are produced,
- how they interact (or not) with one another, and
- what the consequences of these interactions (or separations) are.

It is important to recognize, however, that such questions are asked for a reason – and that the answers that are given have specific consequences for the city itself.

When, in 1937, the urban historian Lewis Mumford asked 'what is a city?', he did
so for a reason and for a purpose. He was writing in the United States of
America at a time when cities there were growing at what seemed to be
alarmingly unmanageable rates. North American cities were sprawling across
the land in a way that Mumford believed to be unplanned and unregulated. For
him, attempts by planners to intervene in this process had been handicapped by
a lack of an understanding of the city. He believed that planners had attempted
to manage urban growth only by tackling the physical structure of cities, mainly
their buildings and roads (see also Boyer, 1983). In opposition, Mumford argued
that planners had not properly understood either the social relations in the city
or the social functions of the city. Although Mumford does not mention these
ideas, he could well have had in mind Ebenezer Howard's diagram of a utopian
city (see Figure 1.5); or Patrick Abercrombie's plan for post-war London (see
Figure 1.6), involving both the dispersal of over one million Londoners into new
towns and the creation of a green belt, where there would be strict controls on
new developments.

ACTIVITY 1.3   Look closely at the plans for the city in Figures 1.5 and 1.6. What
do you think these plans say about how these planners conceived ideal social
relations and how the different functions of the city were to be distributed? ◆

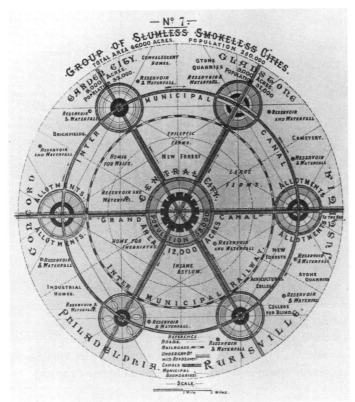

FIGURE 1.5
*A plan of
Ebenezer
Howard's
social city*

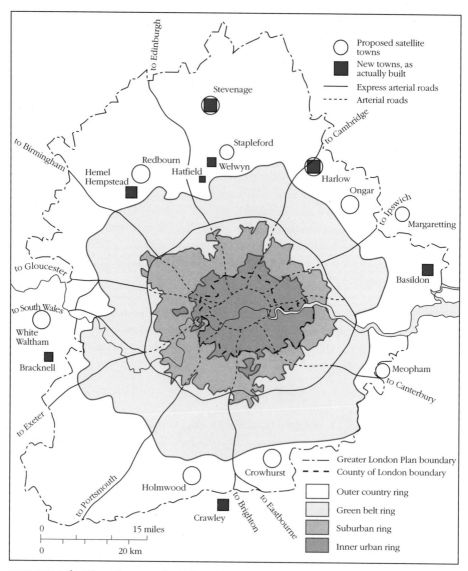

FIGURE 1.6 *The Abercrombie Plan for Greater London, 1944*

One key feature of both these plans is the way in which they tend to separate different aspects of the city, rather than bring them together. This is especially evident in Howard's plan to separate economic activities from one another. In this understanding of social relationships, it was best to locate different aspects of cities well away from one another: classes were to be housed in separate places; while houses were to be built in suburbs, well away from places of work, and so on. Through such plans, urban planners hoped to realize their dream of creating cities with a sustainable environment in which people would live and thrive.

These plans did not just stay on paper, however. Through such schemes, planners' understanding of urban social relationships were rendered concrete, becoming the basis for building new towns both in Britain (famously such as Letchworth, Stevenage, Harlow and Milton Keynes) and also in Europe (in Spain, France, Germany and even communist Russia) and the United States of America (see Hall, 1996; Grace, 1997). However, Mumford was trying to think of the city as more than being a physical fact of buildings or as an array of discordant functions, so he began to think about the city as a *social institution*. What did he mean by this?

For Mumford, one of the best definitions of the city had been given by John Stow – whom Mumford describes as being 'an honest observer of Elizabethan London'. For both John Stow and for Lewis Mumford,

> Men [*sic*] are congregated into cities and commonwealths for honesty and dignity's sake, these shortly be the commodities that do come by cities, commonalities and corporations. First, men by this nearness of conversation are withdrawn from barbarous fixity and force, to certain mildness of manners, and to humanity and justice … Good behaviour is yet called *urbanitas* because it is rather found in cities than elsewhere.

(John Stow; cited by Mumford, 1937, p.184)

We should, of course, quickly note that Stow's observations are innocent of neither class nor gender relations. He was specifically talking about privileged merchant and aristocratic English*men*. For historians there are no doubt intriguing features of his observations, but Mumford is especially interested in what Stow has to say about the character of social relations within the city. In particular, Mumford (following Stow) is suggesting that forms of social interaction occur in cities that are not found anywhere else.

For both Stow and Mumford, city life has an entirely different temperament from places elsewhere. Thus, it is significant both that people can form themselves into 'commonalities' and that the 'nearness of conversation' ensures city-dwellers are inclined towards justice, honesty, dignity and humanity: that is, towards *urbanitas*. Nowadays, we might feel that the idea that the city is characterized by a mildness of manners is quite alien. To begin with, such remarks downplay not only the heightened intensity of urban life, but also the inequalities (in this case, between classes and genders) and conflicts that often scar the city. Perhaps these aspects were ignored by Stow because he was only considering 'polite society'. Nevertheless, it is possible to see in these remarks something about the distinctiveness of the city.

Taking his cue from Stow's argument about the consequences of people 'congregating' in cities, Mumford argues that the city – by bringing people (and their money) together – both enables new forms of association to be created (in Stow's time, corporations, commonwealths and so on), and also requires of people that they interact in new kinds of ways. With this in mind, Mumford attempts his own definition of the city:

> The essential physical means of a city's existence are the fixed site, the durable shelter, the permanent facilities for assembly, interchange, and storage; the essential social means are the social division of labor, which serves not merely the economic life but the cultural process. The city in its complete sense, then, is a geographic plexus, an economic organization, an institutional process, a theater of social action, and an aesthetic symbol of collective unity. The city fosters art and is art; the city creates the theater and *is* the theater. It is in the city, the city as theater, that man's [*sic*] more purposive activities are focused, and work out, through conflicting and co-operating personalities, events, groups into more significant culminations.

> (Mumford, 1937, p.185)

ACTIVITY 1.4   Take some time to think about this quote and list the features of a city that Mumford picks out. Try comparing Mumford's list with the one you produced for Activity 1.1. ◆

For me, the most curious phrase in this quotation is the idea that the city is a *geographic plexus*. The term 'plexus' is derived from anatomy and it is used to describe the networks (plexuses) of nerves, of blood vessels, of tubes for air and food, and so on, that make up animal bodies. So, by geographic plexus, I take Mumford to be saying that the city is made up of many networks through which flows, interchanges and interactions take place (literally). Thus, the city is like a body, living on its different functions: from manufacturing and assembling, to warehousing and storage, to sheltering and domestic bliss, to personality clashes and political intrigue. These functions of the city have both identifiable geographic locations, and also sets of networks (plexuses) which sustain them. Thus, just as the animal body is kept going by bringing in and using the essentials of life (such as air and food) that exist outside the body, urban networks stretch well beyond the confines of the city to bring in and circulate the things it needs. However, we must not get too carried away with the analogy between the body and the city. Let us think more about the city by itself.

Through multiform *networks*, people, commodities, money, and so on, are continually moving through the city, but they are also meeting up in specific places in the city, whether it is the supermarket, the office-block, the home – or wherever. To summarize, Mumford is saying:

- that the city has a distinct physical form
- that this physical form is based on social exchanges of various kinds (economic, institutional, cultural, and so on)

- that these social exchanges are predicated on specific networks, and
- that these networks are *geographical* in at least two senses: first, that they intersect within the city in particular locations; and, second, that they stretch out beyond the city to other locations in specific ways.

However, this is not all that Mumford is suggesting is characteristic of urban life. Such a view of the city does not capture its vibrancy and creativity.

The vast numbers of people in cities also means that there is always the possibility of meeting new people and of undergoing new experiences. As a result, any individual can become opened up to new purposes in life and even to becoming a different kind of person. The city may be a personal drama, but it is also a social drama. The sheer quantity of possible social interactions means that the city becomes a stage for all kinds of stories. Mumford's main point, then, is that the city *like nowhere else* brings people together, into a narrative that is simultaneously personal and social. More than this, the city *intensifies* and *focuses* these interactions, like a magnifying glass concentrating the rays of the sun onto a small patch of ground. The city is a cauldron – hot, ready to catch fire, to burst into flame. Of course, flames can be both beautiful and deadly. So, the theatre of the city can as easily stage stories about mildness and humanity as about conflict and disharmony.

Though Stow had observed that Elizabethan London was a place where people become mild and well-mannered, Mumford was beginning to notice that the ever-sprawling US city was producing personal and social disintegration. For Mumford, the equation was simple: as cities become larger, they cease to be a place of opportunities and creativity; the larger cities become, the more anonymous social interactions become – because city-dwellers have less and less chance of encountering people whom they might know or be acquainted with or even recognize. As the city life becomes dominated by strangers, there are fewer opportunities to meet people in ways that are creative, productive or supportive. (We will return to this issue in section 3 below.)

For Mumford, the spread of the city meant the inevitable dissipation of its humanity and creativity because people were being disconnected from one another. Cities were no longer places of 'congregation', but areas of 'dissipation'. In Mumford's views the desirability and the sustainability of ever-spreading cities was therefore to be seriously doubted. So, he fervently argued that urban development had to be planned. However, for Mumford, urban planners had to appreciate that the city was socially organized. His conclusion was that plans for the city were to be designed in such a way that the benefits of urban life – its intensity, its creativity and vibrancy, its dramas – were sustained and enhanced.

Rather than take this conclusion as the end of the story, let us take some time to reflect on what other implications can be drawn from Mumford's own answer to the question 'what is a city?'.

First, it is clear that cities have something to do with the kinds of association that form between individuals and groups within the city. They can associate for economic, cultural, social or political reasons – and these might be more or less friendly and more or less antagonistic. Indeed, the city offers the possibility of dissociation from other individuals and groups. Questions, therefore, hang over the ways in which we are to understand urban social relationships. However, from what we have seen so far, urban commentators as disparate in perspective (if not experience, since they were all thinking about North American cities in the twentieth century) as Mumford, Lynch and de Certeau agree that what is distinctive about urban life is its vitality, vibrancy, creativity, novelty and intensity, and so on.

Second, the form of urban associations change with 'nearness' or, to be more precise, with the ways in which social relations are made through proximity and distance, closeness and remoteness. This may seem a little cryptic, but we might think here about the difference between meeting a bank manager face-to-face, when you have known them for over twenty years, and telephone-banking, when you'll never meet the person you're dealing with – and, indeed, you may never be in the same country, let alone the same city (see Graham and Marvin, 1995). The meaning and significance of these urban spatial relationships between people within the city are not always self-evident: being able to deal with someone anonymously can be as comforting as it can be terrifying – depending on who you are, where you are. Questions remain, therefore, about how precisely spatial relationships within and beyond the city work (or not). The city, moreover, seems to be unique in the way it concentrates 'things' (such as people, money, information, buildings) spatially, though why and where these concentrations take place also needs further consideration.

Third, it is not just individuals and groups that are significant, but also the ways in which urban institutions operate. These institutions can include organizations such as business corporations, local government and state departments, but we can also think here of bingo-halls, soup-kitchens, churches and the like. The city is not simply a collection of individuals who are reliant on each other in purely abstract ways. Nor is the city simply a stage on which social dramas are enacted. Instead, the city brings people together in particular ways – for specific purposes, within particular contexts – and an appreciation of this is vital if cities are to be thoroughly understood.

Looking back over these three conclusions, we can see that they are not really conclusions at all. Instead of completing our search for the answer to what a city is, each point has raised a whole series of intriguing issues.

ACTIVITY 1.5 Look back over Mumford's conclusions and the three other points that have been made. What other questions can you now identify? My list begins like this:

- How do people form associations (or not) with each other?
- How do proximity and remoteness change the ways in which people relate to one another? ◆

For me, there are many other questions and there is no simple answer to the question 'what is a city?' precisely because of this. Nevertheless, we can begin to see that understanding cities involves:

- the ways in which people 'congregate' in and build cities through their endeavours, as individuals or in groups and institutions of various kinds
- the ways in which the city is a geographic plexus, acting as a focus for exchanges of all different kinds, and
- the ways in which the city becomes a way of life, a social drama that plays out differently for different people.

We have also seen that cities are distinguished by their scale and intensity. These ideas still seem quite abstract, but they will help us in our understanding of cities. However, we need to appreciate how these ideas are related. To achieve this, we will explore these relationships through a case-study. Let us consider the development of a particular US city to see what it can tell us about what the city is like. We will use the example of Chicago, Illinois. But a word of caution before we proceed: it is important to remember that this case-study has been chosen because it sheds light on aspects of our question 'what is a city?', and not because it is typical of all urban development, no matter where the city is in history or in the world.

2 *From nature to metropolis (and back again)*

2.1   CHI-GOUG BEFORE CHICAGO: LAND AND NATURE

By the late seventeenth century, a small settlement had grown up next to a small harbour by the side of Lake Michigan, one of the great lakes in North America (see Figure 1.7). This settlement had first been established in the 1770s for fur trading by Jean Baptiste Point du Sable (see Zorbaugh, 1929/1983). It took its name from a kind of wild onion that the First Nation Americans called *chi-goug* (see Cronon, 1991). For the next half century or so, First Nation Americans (such as the Potawatomis, the Sacs, the Foxes, the Ottawas and the Chipewas) and settlers (such as the English, the French, the Russians and European Americans) came to Chicago to trade the essentials of life – such as foodstuffs (corn, dried meat, flour, fish and alcohol etc.) – for high-value goods (such as furs and jewellery). At that time, the place was quite unremarkable. A small town of a few hundred people, many exchanging goods – goods which were being brought relatively small distances in canoes and shipped longer distances to markets out East (in New York and other eastern cities). Many cultures, many languages, many goods and many people met on the banks of the Lake, but their mostly peaceful interchanges would not last forever.

FIGURE 1.7   *Henry Rowe Schoolcraft's view of Chicago in 1820*

ACTIVITY 1.6   Compare Figure 1.1 and Figure 1.7. Do you think Chicago, at this time, fits your image of what a city is like? Take some time to think about the reasons for making your decision. ◆

For me, 1820s' Chicago looks just like a dispersed village and, furthermore, there seems to be nothing in its built form (or the activities on the waterfront) that suggests that it will become a sprawling metropolis in less than a century. Yet, this is also a place where diverse people come together and meet and get involved in economic exchanges, *just as they might in cities*. So, why isn't Chicago a city? We can conclude, from what we have learnt so far, that Chicago is not yet a city because it does not yet have the size, density and intensity that cities have.

While friction between the First Nation Americans and the settlers had occasionally flared, life in Chicago remained relatively stable. Mainly, this appears to be because the traders (whatever their background) were much more interested in the deals and exchanges they were making, than in warfare and territorial gain. No-one seemed to be exploiting anyone, at least not to extremes. This illusion was to be shattered, however, when the Sac chief Black Hawk – along with a group of many hundreds, comprising Sac, Fox and Kickapoo – attempted for the last time to defend the land from colonization by the white invaders. When Black Hawk was defeated at the Battle of Blood Axe on 2 August 1832, even he could hardly grasp the enormity of what would happen.

Though he was fighting to save the land from white settlement, Black Hawk's beliefs prevented him from seeing how important land ownership would become. The earth, surely, could not belong to anyone. The white colonizers knew otherwise: not only did they know how much money there was in buying and selling land for profit, but they also knew that the faster a settlement grew, the greater the profits would be. If they bought in the right place at the right time, land speculators could become millionaires. If, that is, the village grew into a city. First, however, the land had to be made into property. There was an obstacle: it belonged to the Potawatomis. Within months of Black Hawk's defeat, however, the Potawatomis – who had not actually taken part in the uprising – were quickly forced to give up their rights to the land (as elsewhere: see Cronon, 1983). Now the land could become a commodity, ripe for speculation. The land was divided into plots, which were then sold to speculators.

Chicago began to offer new opportunities for enrichment, if (and only if) the dream of an urban future could become a reality (see Cronon, 1991, Ch. 1). Nothing, however, guaranteed Chicago's development into a great metropolis. Nevertheless, from 1833 onwards, plots of urban land were drawn onto maps, even though the city itself existed only in people's greedy imaginations: 'Fictive lots on fictive streets in fictive towns became the basis for thousands of transactions whose only justification was a dubious idea expressed on a overly optimistic map' (Cronon, 1991, p.32; see Figure 1.8). Chicago, however, was only one of many paper cities, one of many dream cities in which fortunes were made and lost on the basis of optimistic transactions, where speculators effectively gambled on the prospect that a prosperous city would be built on land consisting of little more than mud and tall prairie grass.

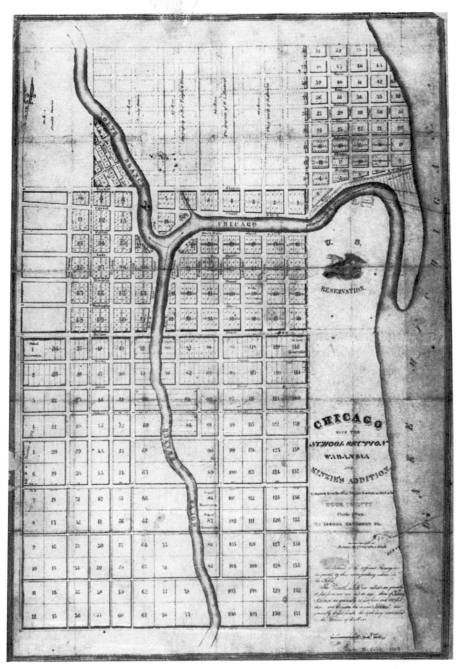

FIGURE 1.8 *Joshua Hathaway's plot map of Chicago in 1834*

Nevertheless, Chicago did have more going for it than some other speculative cities. To begin with, it had one of the few harbours on the Great Lakes. This meant that ships taking goods to and from the cities on America's eastern seaboard had little choice but to use Chicago as a port. Consequently, farmers wishing to get the best price for their products would send them to Chicago, because their produce would ultimately be sold in the richer markets of New York or Boston. Despite this small boon of nature, Chicago still could not grow until it solved a problem set by nature: the weather.

At certain times of the year, travelling around Chicago was effectively impossible. During the rainy season, the roads into Chicago turned to mud and goods wagons could not pass. Even the streets in Chicago became impassable. And, during the winter, the snows blocked the roads. As if this were not bad enough, during the dry season, there was too little water in the harbour for the ships to dock. Sometimes it was impossible for farmers to get their produce to Chicago and, at other times, it was impossible to get it out of Chicago. Only when it was not too dry and not too wet, and only when it was not too hot and not too cold, could Chicago trade. Furthermore, the snow and the mud also limited the geographical extent of Chicago's hinterland. Farmers could only trade in Chicago if they could get their goods there within a certain time-frame – and it would often take them weeks to get their produce there. Chicago's development was hindered: by the seasons, by geography, by land and water.

If Mumford is right to think of a city as a geographic plexus, then, for me, this Chicago barely qualifies – or, at least, it qualifies only at certain times of the year. Unless Chicago could extend its transport networks to draw in goods from a wider area, it could not grow; nor could it grow if people within the city could not move around. Neither the city's streets, nor the roads than linked Chicago to its hinterland, were good enough. The speculators still dreamt of the city's greatness (and the profits they would reap on rising land prices), so solutions were sought in new transport technologies. In this way, had they known it at the time, Chicago would be provided with the infrastructure it needed to become a geographic plexus, the centre for exchanges of all kinds; a place for people to congregate; a city.

2.2   THE MAKING OF CHICAGO: FROM MUD TO MOVEMENT

As the seasons unfolded, Chicago came under siege from nature – but Chicagoans were plotting to break free. Within the city, a quite remarkable plan was adopted. They decided to lift the city literally out of the mud (see Cronon, 1991, pp.55–63). So, from 1848 onwards, the street levels were raised between 4 and 14 feet. In some cases, large buildings were lifted by physical force to their new street level or moved, intact, to alternative locations! Imagine seeing an office-block inching its way down the street, carefully (and fearfully) attended by hundreds of construction workers! It took about twenty years, but eventually it became possible to move around the city at any time of year. While there were improvements in the circulation of people and goods within the city, 1848 also marked the beginning of another innovation: the use of railways. Construction of the first railroad began in March 1848 and, despite difficulties, others quickly followed.

These developments in the transport network both within and beyond the city changed the way in which people and goods travelled in the region. No longer did Chicago have to wait until the snows cleared for the latest fashions to come from New York and beyond – from Europe, especially Paris; no longer did farmers have to wait until the mud had dried before they could get their goods to Chicago. Chicago began to grow as its transport networks pushed further into the surrounding area, enlarging its hinterland. In thinking about Chicago's development into a city, then, it is important both

● to consider the extension of these networks outwards, and also

● to appreciate their concentration on, and intersection in, one place: Chicago.

Chicago was fast becoming the point of exchange: both for goods flowing east towards New York, the east coast of the United States, and beyond; and also for goods flowing in from Chicago's hinterland. As the railway lines extended, it also became clear that they were centred on Chicago – to Chicago's initial advantage. Thus, according to Cronon, no railway starting in the east went further than Chicago, and nor did any railway starting in the west. To travel across America, you would have to stop off *in* Chicago and change trains. In practice, this would probably involve staying for a time in Chicago and, more than likely, this would cost money (and time), though this would not necessarily be unpleasant: the city offered the jaded traveller many delights, some illicit.

ACTIVTY 1.7 Take a look at Figure 1.9. Notice that the major railroads connect the major cities, but also think about where the networks of minor railroads are found. ◆

Chicago began to grow, then, partly because it lay at the heart of a transport network that extended the city's hinterland ever further; but also because Chicago enabled people and commodities to be switched between modes of transport, between rail and ship, canal and road. It was the railways that had the most profound effects. Most significantly, the train changed people's experiences of space and time. Goods and people could travel further and quicker than they had done. Before the railway, in 1852, it had taken over two weeks to get from New York to Chicago; by 1857, the same journey took two days. In this respect at least, both time and space had become compressed (see Leyshon, 1995).

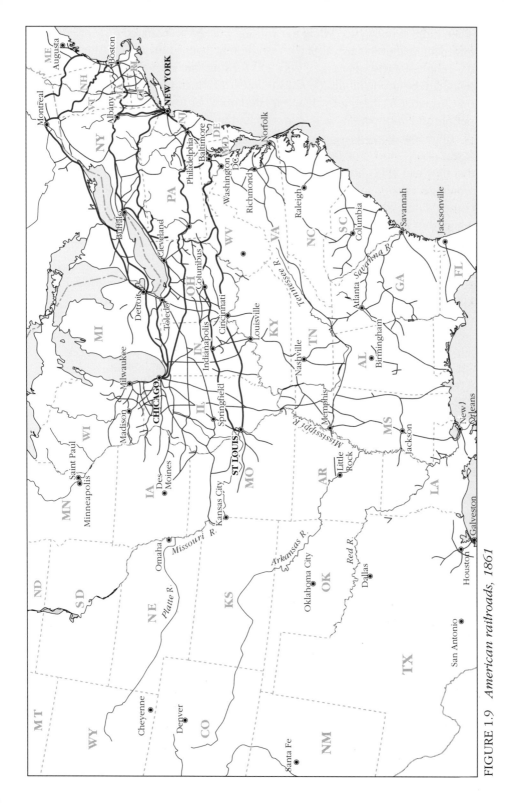

FIGURE 1.9   *American railroads, 1861*

'Even more striking', according to Cronon, 'was the accelerated flow of information after the arrival of the telegraph in 1848' (1991, p.76). Chicago lay at the centre of a vast network of railways, for sure, but since telegraph lines tended to be strung alongside railway lines, information was also concentrated on Chicago. Chicago's growth accelerated both because flows of commodities, people and information, were concentrated on it, and also because these lines of communication were extended further and further into America's Great West. It was not so much that farmers and traders wanted to go to Chicago, it was just that they had little choice: this is where the trains took them; this is where they would get reliable information on the price, demand and availability of commodities, whether selling or buying; this is where the excitement was; this is where they could hear the latest news, see the latest fashions; but here, too, were ever-present dangers, the vices, the tricksters, the possibility of getting lost or trapped or … worse.

Let us briefly review Chicago's progress so far. Chicago was at the hub for many kinds of *connections*: connections which stretched out from the city; connections which drew people together within the city. Through the extension and concentration of these connections, Chicago was integrating, and being integrated with, both the surrounding countryside, and also people in other cities – other cities well beyond the imagination of the original inhabitants; that is, beyond Black Hawk's ability either to save the land from appropriation by white invaders, or to stem the flow of colonists into the area through these wider connections.

By the late 1850s, less than a quarter of a century after Black Hawk's insurgency, the entire geography of Chicago had changed (see Figure 1.10). Changed, not simply because of the bricks and mortar that now cramped the once empty lands, but also because the culture of the place was different. Once delays on the river had been measured in months – as passengers waited for the water levels to rise, as they waited for agricultural produce to fill up the boats – now time was measured in minutes and hours: as messages flashed down telegraph wires, as trains operated to tight schedules. The pace of life had speeded up and the distances covered by flows of goods, people and information were ever greater.

Reflecting on this, we can anticipate that this speeded-up pace of life is characteristically urban because of its *intensity*. But we can also note that this experience of intensity has two aspects. The first has to do with *time and space*. Not only are things moving through Chicago with greater speed, but things are also coming from greater distances. The second has to do with *size*. It is important to note that the intensity of life in Chicago also involves the sheer quantity of things going on. The world was beginning to flow into and out of Chicago, circulating and mixing while it was there. Cronon continues the story:

By fulfilling the role that the railroads had assigned to it – serving as the gateway between east and west – Chicago became the principal wholesale market for the entire mid-continent. Whether breaking up bulk shipments from the east or assembling bulk shipments from the west, it served as the entrepôt – the place in between – connecting eastern markets with vast western resource regions.

(Cronon, 1991, pp.91–2)

FIGURE 1.10   *James Palmatary's bird's-eye view of Chicago in 1857*

27

Here, we can note Cronon's emphasis on the importance of markets in the development of Chicago. However, it is not just the buying and selling goods that is significant. By the 1850s, commodities were being produced and exchanged under specific conditions; conditions which were being increasingly determined by the demand for profit and gain. Thus, the railways signalled a change in the ways in which economic calculations were made in the area. Instead of paying for rail tickets with furs or foodstuffs, people used money. An entirely different form of economic rationality had been set loose: pricing policy, profit margins, accounts, contracts, all began to change the way business was transacted throughout North America. Cronon makes this assessment of the situation:

> The changes that the railroad system initiated would proliferate from Chicago and fundamentally alter much of the American landscape. As the city began to funnel the flow of western trade, the rural West became more and more a part of the hinterland … Wherever the network of rails extended, frontier became hinterland to the cities where rural products entered the marketplace. Areas with limited experience of capitalist exchange suddenly found themselves much more palpably with an economic and social hierarchy created by the geography of capital.

> (Cronon, 1991, p.92)

By the 1860s, through the circulation and investment of capital, through the concentration of production and exchange, Chicago had grown into the great metropolis that the early speculators had dreamt of – only thirty or so years before (see Scranton, 1994). What had happened?

- It was both the brutal disinheritance of First Nation Americans that was significant, and that the land had been converted into property that could be bought and sold as a commodity.

- It was not simply that commodities were traded in Chicago, but that Chicago was the best place to trade: it had a greater variety of goods for purchase and sale; it had better links to other markets and was, therefore, both the most likely place to make a profit and also where the largest profits could be made.

- It was not just that changing technologies had extended Chicago's trade networks, but that Chicago had become a geographic plexus for networks of roads, rails, canals and ships, each funnelling flows of commodities, people, information, capital and so on.

Chicago now lay at the heart of geographic plexuses that stretched across the whole of the American continent, and beyond. But we should note, too, that Chicago's physical existence owed as much to the (capitalist) economic organization of these networks as to its position in these networks. However, we should not be fooled into thinking that Chicago's success (measured in profits and investments) rested only on its ability to draw people into capitalist calculations and practices through its ever-extending networks, for Chicago was also being built out of something much more precious: nature.

2.3   CHICAGO'S NATURE: THE CITY'S FOOTPRINT ON THE FACE OF THE EARTH

When the cultural geographer Yi-Fu Tuan asked 'what is the essential character of the city?', he argued that 'cities are artefacts and worlds of artifice placed at varying distances from human conditions close to nature' (Tuan, 1978, p.1). By also acknowledging that life differs in its intensity and character, Tuan suggested that:

> Cities, then, may be ranked according to how far they depart from farm life, from the agricultural rhythm of peak activity in the warm half of the year, and from the cycle of work during the day and of sleep at night. At one end of the scale we have the village subordinate to nature; at the other, the city that does not know how it is fed, that comes alive in winter and slights the daily course of the sun.

> (Tuan, 1978, p.1)

For me, Tuan is suggesting that cities can be classified by their distance from nature: whether the city is a city is determined by the extent to which it has cut itself loose from the course of the sun and the vagaries of the seasons. If we think about Chicago in the 1840s, we can see that it has hardly got off the ground: indeed, it is mired in the mud and snow. By the 1850s Chicago has lifted itself out of nature – by conquering both winter and night – so that, popularly, 'Chicago represented all that was most unnatural about human life. Crowded and artificial, it was a cancer on an otherwise beautiful landscape' (Cronon, 1991, p.7).

But is it appropriate to see the city, essentially, as distant or separate from nature? In order to answer this question, we can look more closely at Chicago's relation to nature – and, in particular, to trees. What is important is the way in which trees become a marketable (or unmarketable) commodity through their relationship to Chicago, where various demands for wood are focused. First, let us deal with the demand for wood.

As farmers moved into the treeless prairies of the American Mid West, they found that they needed wood: its resilience and malleability made it ideal for building houses, barns, fences; and it was also needed as a fuel for heating and cooking. In the towns, wood was used to make sidewalks. Meanwhile, railroad and shipping companies used wood in the construction of tracks, rolling-stock

FIGURE 1.11 *Chicago's lumber district*

and ships. Across the Mid West lumber was in demand. Though the white pine was ideal material – because of its strength, plasticity and buoyancy – the forests were hundreds of miles to the north and west of the prairies.

Somehow the trees had to travel. They were cut down and turned into logs. Through a dense network of streams, rivers and lakes, they were floated – then shipped – towards Chicago, which quickly developed into a 'lumber city', where vast stocks of lumber were stored in huge yards (see Figure 1.11).

While lumber travelled through Chicago to the Mid West, foodstuffs made the return trip: from the farms of the Mid West, through Chicago, to the forests. Chicago, then, did not just lie on the edge of two different geographies – one of prairies and one of forests – it made them overlap: partly by transferring commodities – food and wood – between them; and, partly, by converting both the forests into lumber and the prairies into grain-producing farms and pig-rearing ranges. Chicago, therefore both lived off the land, and created a new geography by integrating the markets for wood and food across the American Mid West.

The white pine did not chop itself down. The lumber firms required labour: men were used as lumberjacks, but also in the saw-mills, to break up the log-jams that often – and very dangerously – blocked the flow of streams and rivers, and in the processing and selling of the wood; meanwhile, women were used to wash and cook. For all, it was a tough life and it was not made any easier by conditions in Chicago. Since people gathered in Chicago to look for work, competition amongst them meant that wage labour prices were driven down – good for employers, but not so for the employees, who frequently would have to take drastic measures to get what payment they could for their work.

Chicago's importance in the lumber trade depended both on a pool of cheap labour, and on a ready market for wood. Farmers from across the Mid West were arriving in Chicago to sell their goods and they were also looking for wood. For the lumber firms this meant that they could guarantee that they would be able to make a sale, sooner or later. This in turn meant that ships and trains were rarely empty: because the railroads carried food from, and wood and equipment to, the prairies; because ships carried wood from and food and equipment to, the forests; because, in both directions, people and money flowed into and out of Chicago:

> No place was more important in coordinating this massive movement of water, men, and wood than Chicago. The city served as the chief lumber market on Lake Michigan but its role went much further than just buying and selling wood. Many Chicago lumber dealers participated in every phase of regional lumber production, and Chicago capital thus often directed the movement of white pine from forest to mill to final customer.

(Cronon, 1991, p.159)

Chicago became the point of contact and overlap between geographies of wood, food and equipment, of people and money, of government and the regulation of trade, and of railways and waterways. As Chicago grew, flows of commodities became concentrated on it, as they were *channelled* through its transport networks: 'The commodities that flowed across the grasslands and forests of the Great West to reach Chicago did so within an elaborate human network that was at least as important as nature in shaping the region' (Cronon, 1991, p.164).

These commodity flows were of different kinds – not just wood, but also grain and pigs – and each had their own effects on the city's physical form: as timber-yards, huge grain-silos, timber-mills, meat-packing companies and factories for agricultural and forestry machinery sprang up. Moreover, 'City streets became places where the products of different ecosystems, different economies, and different ways of life came together and exchanged places' (Cronon, 1991, p.61).

Chicago's streets and markets thronged: the huge quantities and varieties of goods coming into the city, along with the large numbers of people, were being *concentrated* in the city – and this encouraged more people and more money to try their luck there. Chicago was acting both as a magnet and as a magnifying glass; it pulled in – and exaggerated – greater quantities of commodities and money, people and information, and so on. Chicago had reason to grow: people could live there, work, there, and perhaps make their fortune there.

As Chicago's networks were extended, its commodities came from further and further away – and, as the city grew, its demand for wood increased. As the forests were cut down, lumber operations moved further and further away from the city; for no-one was able to seriously challenge the desirability of chopping the forests down. If you had been able to look down from an aeroplane, it would have looked as if a burglar had left an indelible, giant footprint in the forests; the 'fence', receiving the stolen goods, was the city. However, the idea that a city stamps its footprint on the face of the Earth has more than metaphorical resonance. It also gives us a way of describing and assessing the ecological reach of the city and the environmental consequences of urban development. In this sense, the footprint dramatizes both how, and from where, nature is being sucked into the city's factories and markets, and also how, and to where, the city spews out its effluents, products, smoke, refuse, and so forth (Merchant, 1994, p.136).

Chicago's footprint was growing bigger and bigger: especially as the railway network expanded to catch up with the new frontiers of logging. And the feeding frenzy continued unabated – until there were no more trees.

We can look back over this story and make some observations:

- Chicago lay at the heart of diverse webs of *interconnections* that enabled the overlap of many different markets, commodity flows and resource regions.

- Although Chicago was the point at which these webs intersected, they still needed to be made to work: partly through co-ordination by the firms operating transport linkages (such as the railways companies); partly through the regulation of commodities by Chicago's Board of Trade; and, partly through the regulation of land and resources by the federal government.

- These webs were not innocent of power relationships. This can be seen both in the way that wage labour and trade were organized, and also in the relationship between nature and the city: as nature was brought into the city, converted into commodities, by processing or manufacturing, for resale.

By tracing commodities – such as wood – back to their origins, it is possible to see not only how nature arrives in the city and also how the city is built through nature, but also how far away nature is from the city. In one sense, the city is never distant from nature; as much as it might conquer winter and the night, it does so only by manufacturing itself and nature anew – electric fires and lighting ultimately use nature, in the form of gas, coal, water, nuclear energy and so on. On the other hand, the city also seems to distance itself from nature, as it struggles to 'escape' the environmental consequences of its actions (Merchant, 1994). Nevertheless, nature might have the final word, creating a series of ragged ends for the city. At this point, we can begin to wonder what did happen when the trees ran out.

2.4   THE ENDS OF CHICAGO

Between 1833 and 1848 many people optimistically hoped that Chicago would become a great city. They imagined that, as soon as the canal between Chicago and St Louis was built, trade routes to the southern states would be opened up. However, in this period, the construction of the canal was continually delayed – and Chicago's potential could not be realized. Instead, St Louis dominated the region because of its superior waterway routes. It was, therefore, St Louis that could have become the East–West gateway city. Why didn't it? The answer lies in some speculative decisions that were taken in Chicago.

In 1848 a group of politically and economically powerful people concluded that, without the canal, the future of Chicago rested on a new technology: the railway. No-one, however, had enough money to build it; nor did they have the remotest idea what it took to run a railway (or they might well not have even attempted it?). Nevertheless the Chicago businessmen set up a railway company, but its finances came mainly from very small investments by local farmers, who were promised in return a train station that would link them to Chicago. Only because thousands of farmers believed in the sales-pitch of the railway company, and then decided to make their small, risky investments, did the railway ever get built. Of course, once it was built, it made sense for new railway lines to end up where the existing railways already terminated, that is, in Chicago. Since St Louis simply did not have the connections, it could not achieve the same status. The important point, here, is that Chicago's growth relied both on the development of diverse, wider connections, and also on the way in which these connections were maintained and extended to Chicago's advantage.

ACTIVITY 1.8   Now, think about Chicago as it was in 1820 (see Figure 1.7 above). Why do you think Chicago developed into a city? Think about the changes that have occurred, both in transport and trade, and in the city's physical structure. ◆

An important reason for Chicago's development is that it lay at the centre of many overlapping networks, through which flowed commodities (such as grain, wood and pork), people, information, money, and so on. More than this, warehouses, docks, railroads, factories (and suchlike) sprang up to take advantage of, and to increase, the flow of commodities (see Cronon, 1991, Ch. 4). However, the city's importance does not mean that these networks are necessarily inevitable, permanent or desirable.

- As we have seen, the centring of the railway network on Chicago initially arose out of the tiny, uncertain decisions of thousands of farmers.
- Later, as the railway system expanded, Chicago ceased to lie at the centre of the network as other gateway cities developed.

- And, we should note, the concentration and intensification of commodity flows on Chicago also became a hindrance as they began to congest and endanger the city. In order to exemplify this, let us return to the city's roots: the white pine tree.

As we have seen, Chicago's voracious appetite swallowed up vast areas of forests. Over time, trees came from further and further away. The railway network expanded to meet the needs of the lumber companies, who could no longer float trees down river because the logging frontiers were higher and higher in the mountains, where the streams were too small. The extension of the railway network, however, meant that lumber companies no longer needed to ship their wood to Chicago. Instead, they could sell directly to customers. Chicago was no longer the centre of the railway network and it could be easily by-passed. By the 1880s other cities were growing up over the Mid West, each able to occupy central positions within overlapping networks and commodity flows. As the nineteenth century came to a close, Chicago gradually ceased to be the centre of flows of grain, wood and meat. On the other hand, it was now becoming a great industrial centre (but this is another story: see Cronon, 1994; and also Page and Walker, 1991). The important point to recognize is this:

> All western cities served as markets for their hinterlands, but Chicago did so with greater reach and intensity than any other. By assembling shipments from fields, pastures, and forests into great accumulations of wealth, the city helped convert them into that mysterious thing called capital … As the city's population increased, as its buildings expanded out into the prairies, and as its factories and warehouses spewed forth a seemingly endless stream of goods, so did its capital – which served as the symbolic representation of all these things – continue its preternatural growth.
>
> (Cronon, 1991, pp.148–9)

Chicago's success in the later half of the nineteenth century relied both on its position as a gateway city, and also on manufacturing and warehousing – otherwise, how would Chicagoans have made a living in the city? Chicago's success

- in concentrating flows of goods, people, money, information and so on,
- in building a physical infrastructure of houses, factories, warehouses, depots, and so on,

would, however, also be the source of its problems. It simply became too difficult and too expensive to trade in Chicago.

Having conquered the mud, the concentration of traffic meant that Chicago had become heavily congested by the 1880s. For example, people wishing to travel from New York to Los Angeles would have to spend almost an entire working-day crossing Chicago as they transferred between one railway operator and another. Soon, as the costs of doing business in Chicago rose, it became easier to make profits in other cities.

Just as the railway had once ensured the centralization of flows of commodities, people, money and information on Chicago, now it permitted their dispersal to other cities, such as Minneapolis, Denver, Kansas City and Omaha. Chicago ceased to be the only point of exchange for goods such as meat, grain and lumber. The connections between places had changed as cities grew and increasingly participated in wider national and international economies.

Thinking about what a city is, then, involves considering the ways in which flows become concentrated (or not) on one place, such as Chicago. However, this does not complete the picture. It is important to track these flows

- both as they extend outwards beyond the city
- and as they intertwine or disperse within the city.

For this, we need a geographical imagination that involves both the reach of networks and their direction, and also at the same time can trace these networks within, and their consequences for, the city. This, indeed, is to think more carefully about the city as a geographic plexus, but it means more than this too. For, in this story about Chicago, it can be seen both

- that networks shift and change over time, and
- that networks are the result of specific social relationships (in Chicago's case, involving nascent capitalist calculations of profit and production), involving – more often than it might seem – uncertain, thoughtful, inventive, devious, thoughtless, even sometimes foolhardy, behaviour by individuals and groups.

But is the idea of the city as a 'geographic plexus' enough to understand the social patterning of the city? So far, the city of Chicago has been treated almost as if it was undifferentiated – that is, as if Chicago itself was the same all over. However, Chicago itself had (and still has) many worlds within it.

2.5   COSMOPOLITAN CHICAGO

Into Chicago poured thousands of people, from all over the world. Uprooted from their homes, their cultures, they began to make new homes, to recreate their cultures, in this unfamiliar environment. As people came to settle in Chicago, the city grew. This growth involved extending Chicago's networks, through which flowed resources, commodities, people, money, information, and so on. It involved, also, the creation of an industrial base, which provided (or took away) employment (see Scranton, 1994; Page and Walker, 1994). And it involved building ever more (wooden) houses on the grassy prairies. The expansion of Chicago, however, had consequences for Chicago itself.

In this section we will look more closely at the social patterns within Chicago. The evidence that we will use is drawn from work conducted by Robert Park, Ernest Burgess, Harvey Zorbaugh, Louis Wirth and others in the Chicago School of Sociology in the 1920s and 1930s. This is helpful for us, not because it provides the conceptual or analytic tools (mostly because these have severe flaws: see Smith, 1988, and Savage and Warde, 1993), but because they undertook many detailed studies of Chicago's social patterns. In this regard, what is of interest here is the way in which they described both the social and spatial organization of Chicago in the 1920s (and not their views on the underlying principles governing the internal growth of the city). This reinterpretation of their work will allow us to identify some further characteristics of the city, relating to its cosmopolitanism.

One device which Ernest Burgess (1925/1984) used to describe the social–spatial organization of Chicago (and all US cities) was a map: see Figure 1.12.

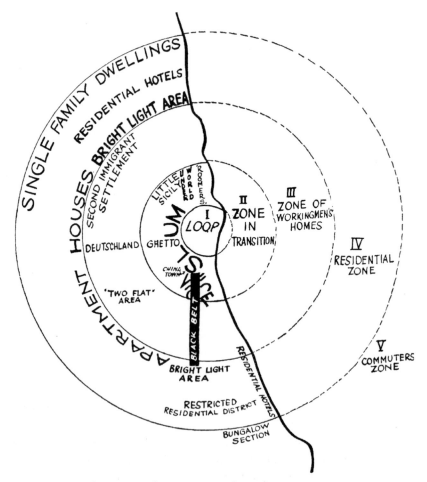

FIGURE 1.12   *Urban areas from Burgess (1925)*

ACTIVITY 1.9   Look at Figure 1.12. The thick line running north–south marks the shoreline, with Chicago to the left, and Lake Michigan to the right. First, study the various concentric rings that divide the city into concentric zones. Now, concentrate on the various districts of Chicago, such as Little Sicily and the Black Belt. Now, what do you notice when you compare these two social patterns of Chicago? Think here about the ways in which the zones and the districts coincide or overlap – or not. ◆

For me, in Burgess's map, two contrasting ideas about Chicago's social patterns have been overlain: on the one hand, there is a series of concentric circles; and, on the other hand, a number of distinct districts have also been identified. These ideas sit somewhat uneasily with one another. Let us see why.

Let us look first at the concentric circles. For Burgess, these *concentric zones* were universal features of all cities (and not just Chicago). He argued that each zone had a specific character:

- Zone I represents the central business district; in Chicago this area is called 'The Loop'.

- Beyond Zone I is a zone in transition. Zone II is occupied by the poor, living in slum conditions. However, this area is 'invaded by business and light manufacture', as represented by the dotted line that marks the factory zone. This invasion by factories causes a displacement of the poor residents into outer zones.

- In Zone III live people who worked in the industries, including those who have escaped from the zone in transition. Zone III is not, therefore, what we would commonly call the inner city, but a richer area where people can afford to live away from their place of work.

- Beyond this zone is a residential area (Zone IV), with high-class apartments and exclusive areas of single family dwellings.

- And beyond that, beyond the city limits, lies a commuter belt (Zone V), in which lie suburban areas and also satellite cities.

Although this model of the city seems static, it also incorporates the idea that people move outwards from the centre of the city as they get richer. Each movement outwards causes further displacements, as richer people vacate deteriorating houses and poorer people move in to occupy these houses.

On the other hand, these zones only really work on the placid waters of Lake Michigan. Once they strike the land, all kinds of complexities seem to interfere with them. There have been various criticisms of Burgess's concentric zone model (Savage and Warde, 1993). We can note, for example, the following issues. The concentric zones do not describe the social (class) patterns in other US cities (see Davies and Herbert, 1993). Not only were rich people choosing to live in élite sectors near to the centre of cities, even in close proximity to slums (curiously, also a feature of Chicago in the 1920s, as we will see), but also the poor were frequently found in peripheral areas. Nor did the model apply well where governments had intervened significantly in the land and housing market: as in centrally planned countries, such as the former Soviet Union; or, through social housing and welfare provision, as in the UK. These criticisms are less important, however, than considering more carefully the other aspect of Burgess's social map of Chicago, that is, its districts. While the concentric zones seem to be flat surfaces characterized both by their internal uniformity and by their degree of difference from each other, the districts of Chicago seem to cross-cut and emboss these surfaces, giving a different sense of the social surfaces of the city.

Let us look again at Burgess's map of Chicago (Figure 1.12). Chicago's concentric zones are clearly overlain by other kinds of social area, predominantly characterized by the presumed ethnicity or 'race' of their inhabitants. There is, for example, a 'Black Belt', a 'Chinatown' and a 'Deutschland'. We can question whether these fit well into the concentric zone model: are all Germans workers; are all Sicilians in slums; why does the 'Black Belt' cross three zones? These questions are important because they relate:

- both to the ways in which people are distributed – by class, by 'race', by gender and by all kinds of other differences – through the city
- and about how these differences between, and amongst, people are interrelated – or not.

Let us look, therefore, more closely at the area in Zone II to the north of 'The Loop'. This area, known as the Near North Side, has a high concentration of different districts, according to Burgess. In it are districts labelled as 'Little Sicily', 'Underworld' and 'Roomers'. In his classic study, *The Gold Coast and the Slum* (1929/1983), Harvey Zorbaugh describes these districts in fascinating, and sometimes poignant, detail. From this study, we can surmise that the unnamed district to the east of 'Roomers' is the 'Gold Coast'. Let us look at Chicago's Near North Side (see Figure 1.13), an area only 1.5 miles square – a mere thirty-minute walk, north to south, or east to west (see Figure 1.14).

In the 1920s the Gold Coast (as you might anticipate) was characterized by luxurious apartments and expensive hotels (and still is). Here lived the rich, with their swanky cars; their forbidding mansions; their extravagant lifestyles, splattered all over the society pages of the newspapers. When the summer was hot, according to Zorbaugh, the people from the slums would look up from the lakeside beach, where they could see the splendour and glamour – but the rich would only heed the poor when they read about murders, violence and social unrest in the slums. They might as well be on different planets: though they lived in close spatial proximity to one another, their social distance could not be greater.

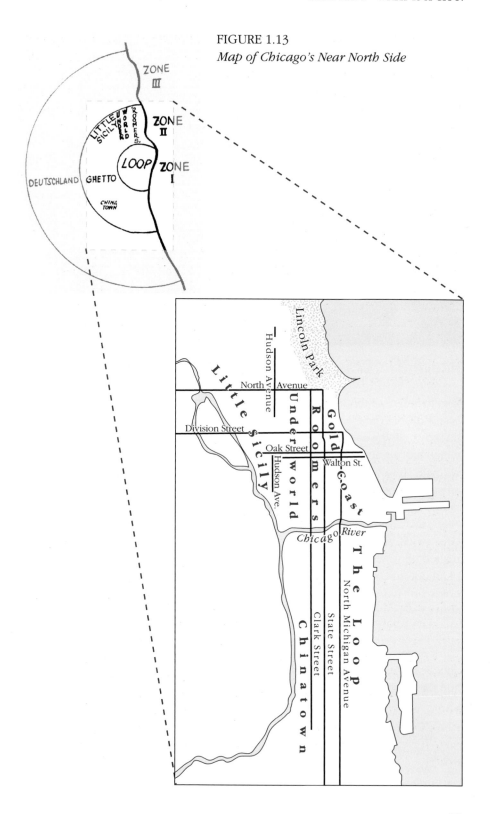

FIGURE 1.13
Map of Chicago's Near North Side

FIGURE 1.14   *Chicago's Near North Side, 1995*

Less than five or ten minutes walk away from the Gold Coast was the rooming-house district (labelled 'Roomers' on Burgess's map). This area, perhaps unexpectedly, was described by Zorbaugh as bohemian. It was a poor area, which had attracted younger, adventurous, single people – most of whom were women. Here were to be found stores selling books, art, antiques and curios; artists' studios; radical clubs; small theatres; and tea-rooms. (Were any of these on the list of urban features you produced for Activity 1.1?) Like many other US cities, this area was known as 'the village' (like New York's more famous Greenwich Village). Intriguingly, it was deserted during the day, as people rushed off to work – as artists, shop assistants (mainly women), inventors, 'men of affairs' and 'women of fashion', waitresses, clerks and entertainers. What was noteworthy, argued Zorbaugh (anticipating Mumford's view, see section 1.2 above), was the great variety of opportunities that large cities offer for these people, even though their livelihoods and social lives remained precarious.

Clark Street, running north/south, divides the Near North Side in two. This is the area described as 'Underworld' on Burgess's map. In both parts, crime is rife – and the street itself is notorious. Zorbaugh describes the scene:

> Clark Street is the Rialto of the slum. Deteriorated store buildings, cheap dance halls and movies, cabarets and doubtful hotels, missions, 'flops', pawnshops and second-hand stores, innumerable restaurants, soft-drink parlors and 'fellowship' saloons, where men sit about and talk, and which are hangouts for criminal gangs that live back in the slum, fence at

pawnshops, and consort with the transient prostitutes ... It is an all-night street, a street upon which one meets all the varied types that go to make up the slum.

(Zorbaugh, 1929/1983, pp.10–11)

The slum is not solely inhabited by Italians – as you might have expected given that it is labelled 'Little Sicily'! In fact, there are many migrant groups: Poles, Irish, Slavs, African-Americans, Persians, Greeks, Germans (many of whom were Jews) and over twenty other nationalities. Though living in close proximity in the crowded blocks to the west of Clark Street, these groups would pass by or cross over one another, rarely meeting in any significant way; rather they tolerate each other's presence with a kind of indifference. Even so, the slum is a dangerous place. Zorbaugh, for example, notes the mob-related murders at the corner of Oak Street and Cambridge Avenue, known locally as Death Corner; and a brawl between 200 Syrian and Assyrian Persians at a coffee shop on Clark Street, because no Syrian Persian was allowed north of Huron Street. Many different worlds were brought together in this densely populated part of the city. These worlds had their own horizons, back to the villages dotted across the Mediterranean: in Sicily, Greece, the Middle East, Eastern Europe, and so on. Sometimes these worlds collide, other times they simply slide past one another, as if they weren't there.

We have taken a little time to think about the *cosmopolitanism* of Chicago. However, we should also note that this cosmopolitanism does not emerge from nowhere. Migrants to Chicago arrived through many different routes and ended up in distinctive parts of the city. For example, black migrants had been arriving in Chicago from the American South since the end of the Civil War in 1865. Likewise, Germans, Italians, Poles, Slavs, Greeks and Persians had been leaving the most impoverished parts of Europe since the end of the First World War in 1918. Implicitly, therefore, Burgess's map of Chicago's social and spatial organization also evokes the different histories and geographies of migration into the city – and, more than this, the social dramas of people living in Chicago.

1920s' Chicago is a *city of contrasts* – between different ethnicities, different lifestyles, different classes, and so on – and these contrasts exist side by side. But there is no reason why this should be a problem. Except that Chicago is also a *city of exaggerations*: it concentrates and intensifies these contrasts. In Chicago these exaggerations seem to make it a bundle of nerves, jangled by crime, by 'racial' and communal tensions, by excitement and pleasure, by social and political unrest, and by unending influxes of people. For the Chicago School, this suggests that city life has an intensity – a way of life – not found elsewhere. In our exploration of the question 'what is a city?', we have occasionally touched on the idea that what distinguishes city life is its *intensity*: its speed, its breadth; its opportunities, its anxieties; its scale, its social interactions. It is to this key aspect that we turn next.

3 *The intensity of city life: size, density and heterogeneity*

From the arguments presented so far and as exemplified in the case-study of Chicago, it has been shown that cities are more than simple assemblages of physical features, more than simply places where lines of communication intersect, more than simply a gateway for flows of goods, people, money, information, and so on (see Castells, 1990; 1996). It is not that these are unimportant. It is, rather, that cities add something more:

> The city ... is something more than a congeries of individual men [*sic*] and of social conveniences – streets, buildings, electric lights, tramways, and telephones, etc.; something more, also, than a mere constellations of institutions and administrative devices – courts, hospitals, schools, police and civil functionaries of various sorts. The city is, rather, a state of mind, a body of customs and traditions, and of the organized attitudes and sentiments that inhere in these customs and are transmitted with this tradition. The city is not, in other words, merely a physical mechanism and an artificial construction. It is involved in the vital processes of the people who compose it; it is a product of nature, and particularly of human nature.

(Park, 1925/1984, p.1)

Like Lewis Mumford, Robert Park was not impressed by the idea that the city could be defined by its physical form. Whatever it was that makes a city a city, it had more to do with its social processes. The city does not just express itself in the buildings, the streets, the traffic that seem to define it, but in the ways in which people live, work, trade and enjoy themselves – *or not*. Nor is it simply a social institution that involves the courts, hospitals, schools, the police, bureaucracies and city government. What is vital about cities is that they bring together people in such a way that this makes a difference to what goes on between them. What makes the city a city is not only the skyscrapers or the shops or the communication networks, but also that people in such places are forced to *behave* in *urban* ways: this is where the 'rat race' is; the city buzzes and roars – and it never sleeps.

According to this interpretation of the city, there is something distinctive about the intensity of social interactions in the city. This is defined by their speed, their heterogeneity and, perhaps paradoxically, the ambivalence in people's experiences of the city. However, this description requires some explanation. Why might living in the city exaggerate size, contrasts, heterogeneity, ambivalence, speed and so on? Addressing this question will tell us something more about what a city is.

In 1938 Louis Wirth attempted his own definition of the city: 'For sociological purposes a city may be defined as a relatively large, dense, and permanent settlement of socially heterogeneous individuals ...' (Wirth, 1938, p.190). It is not so much the accuracy of this definition that is important, but the three characteristics that Wirth chooses to emphasize. The city has:

- large numbers of people
- a density of settlement, and
- a heterogeneity both of individuals and of group life.

Each of these characteristics has ramifications for the ways in which people in cities come together (or not). Let us deal with each in turn.

3.1   THE NUMBER OF PEOPLE

To begin with, Wirth is suggesting that the large numbers of people in cities makes a difference to how people interact (or not) with one another. There are three basic reasons for this:

- because of the greater range of variation between individuals
- because of the greater numbers of social interactions and
- because of the greater potential for differentiation amongst people.

With vast numbers of people living in the city, there are bound to be wide range of variations amongst them. According to Wirth, this '*should* give rise to the spatial segregation of individuals according to colour, ethnic heritage, economic and social status, tastes and preferences ...' (1938, p.191, emphasis added). To this list, we can usefully add others, such as gender, sexuality, age, able-bodiedness, and so on. More importantly, Wirth does not explain why spatial segregation *should* arise from variation between individuals, by 'race', class, or whatever. Therefore, he does not describe the social processes underlying either spatial segregation or social differentiation, and so Wirth cannot see the kinds of prejudices and inequalities that might ensure that people are kept at a distance.

Nevertheless, Wirth does elaborate on the ways in which large numbers permit, or force, people to form new kinds of *social interaction*. He argues that people in cities have more opportunities to form bonds that do not rely on kinship ties, neighbourliness, communal sentiments, tradition and 'folk' attitudes. This is because there are other grounds for friendships between individuals and solidarity amongst groups. At this point, we can remind ourselves of John Stow's remarks (page 15 above). Remember, he talked of the city's 'commonalities and corporations'. Perhaps the possibility of forming new associations is reason enough to live in the city. On the other hand, the 'nearness of conversation' with others might not necessarily lead to 'a certain mildness of manners', as Stow suggests. For Wirth,

> The contacts of the city may indeed be face to face, but they are nevertheless impersonal, superficial, transitory, and segmental. The reserve, the indifference, and the blasé outlook which urbanites manifest in their relationships may thus be regarded as devices for immunizing themselves against the personal claims and expectations of others.

(Wirth, 1938, p.192)

Wirth is describing encounters between people in cities as impersonal, superficial, transitory and segmental, while city-dwellers themselves are reserved, indifferent and blasé – and necessarily so. Part of the reason for this is the sheer number of people with whom city-dwellers have to interact: it is just not possible to get to know everyone. For Wirth, then, interactions within cities are characterized by a kind of superficiality and anonymity. So, is this a good thing or not? Wirth is ambivalent; he sees in this situation both a good side and a bad side:

- on the one hand, because contacts are superficial and anonymous, individuals are released from any obligations or expectations that might arise from living within a tightly controlled community;

- on the other hand, while urbanites are emancipated from traditions, ties and constraints, they also forfeit a sense of participation in communal life and sacrifice the capacity to relate to other people as if they were part of a community.

For Wirth, the city therefore *disorganizes* social life. By this, he means that previous forms of social organization are undermined and replaced by social relations whose (unorganized) organization is based on *indifference*, superficiality and a kind of utilitarianism in which everyone is out for their own interests. But how definitive is this of cities in general? Thinking about when and where Wirth was writing, it is not difficult to imagine this description of US cities at the time of the Great Depression as being largely accurate, at least for many groups arriving and living in cities. Nevertheless, we should be careful about seeing large numbers as producing only indifference or superficiality or utilitarianism.

From Wirth's analysis, it is possible to draw another conclusion. Instead, we can glimpse something of the paradoxes of city life: its tensions, its ambivalences. For example, there is something paradoxical here about the individual's relationship with city life: it is both liberating and stifling; both stimulating and deadening. Rather than saying that all urban life is like this or like that, it might be better to consider the ambivalent or paradoxical experiences that cities offer – and how these will be different for people in different social and geographical locations in the city. It may be, therefore, that what is characteristically urban is that it is paradoxical; paradoxical, both in the sense that it embodies elements that are seemingly opposed at one and the same time, and in the sense that the seemingly opposed elements are brought together, intensified and concentrated, in the city.

3.2 THE DENSITY OF SETTLEMENT

In Wirth's sociological definition of the city, density of settlement was the second fundamental characteristic. But why is density important? Wirth is less concerned with the way in which the city concentrates people into a limited space (thereby producing density of settlement), than with the effects of this compacting on people's way of life. Drawing on Simmel (1903/1995), Wirth suggests that density leads to

> ... the close physical contact of numerous individuals [which] necessarily produces a shift in the medium though which we orient ourselves to the urban milieu ... our physical contacts are close but our social contacts are distant ... We tend to acquire and develop sensitivity to a world of artefacts and become progressively further removed from the world of nature.

(Wirth, 1938, p.192)

In this account, the concentration of people – which has already led to superficiality and indifference – also produces forms of distancing, not only from other people, but also from nature. People become indifferent to everything, that is, except the world of artefacts. In this way, people become indifferent to the 'glaring contrasts between splendour and squalor, between riches and poverty, intelligence and ignorance, order and chaos' (Wirth, 1938, p.192). Once more, the city is seen as characteristically paradoxical: people live in close proximity to others, yet they don't care about them; city-dwellers develop an insensitivity to others, yet are highly sensitive to the value of artefacts. But is this process inevitable? Clearly not. Why? Because the city contains, paradoxically, both the possibility for individuals to differentiate themselves from others, and also the opportunity for like-minded individuals to form new associations:

- on the one hand (as we have seen, in both sections 1.2 and 3.1), individuals are freed from social obligations and ties and this allows them to become different people;
- on the other hand, the concentration of different people in cities means that individuals can readily find others who are like them – and, in groups, they can produce areas of the city in their own image.

ACTIVITY 1.10 Think about Chicago's Near North Side (section 2.5). Here, there was a concentration of very different groups. Think about the different kinds of people in the area. Where did they live? Did they interact with one another or not? ◆

For me, Chicago demonstrates that communal ties can be formed in cities and that people from similar backgrounds tend to live in close proximity. Paradoxically, this does not mean that they necessarily interact with one another, nor does it prevent very different, even antagonistic, groups from living (almost) side-by-side. For Wirth,

> The city consequently tends to resemble a mosaic of social worlds in which the transition from one to the other is abrupt. The juxtaposition of divergent personalities and modes of life tends to produce a relativistic perspective and a sense of toleration of differences which may be regarded as prerequisites for rationality and which lead toward the secularization of life.

(Wirth, 1938, p.193)

Wirth is arguing that the *coexistence of differences* within the city produces greater tolerance amongst city-dwellers – although not without the occasional bitter conflict. In this, Wirth sees the potential for an urban way of life that is rational and secularized. For him, these represent positive possibilities for city life.

However, there is a negative side too. Wirth recognizes that people from different social worlds, as reflected in inequalities and prejudices, can be vulnerable to exploitation and exclusion. This can be seen in the *differentiation of urban space*. For example, desirable areas of the city (like Chicago's Gold Coast) exclude those who cannot afford to live there; while slum districts (such as Little Sicily or the Black Belt) become traps for stigmatized people – such as slum-dwellers, black people, immigrants, women, and so on – who cannot escape because they are excluded from better housing and employment opportunities. There is another negative consequence of density: the greater the number of social interactions, the greater the possibility that they will interfere – or conflict – with one another. Therefore, he argues, there needs to be a degree of co-ordination of everyday life, necessitating new forms of control. These, he suggested, were epitomized by two features of the city – the clock and the traffic-signal. (Were these in the list you produced for Activity 1.1?)

Both the clock and the traffic-signal control the routines – the ebb and flow – of urban life. But what is it about city life that needs such control? For Wirth, both the sheer quantity of people and the concentration of people produces a speeded-up pace of urban life. City life seems to be faster

- both because things move more quickly in the city (and if you have ever been grid-locked, or frozen at a bus-stop, then you'll know this is not always true!)
- and because of the amount of things that are going on
- and also because city life is more extensive – extensive in the sense that elements are brought together in the city through many different networks, which stretch out within and/or beyond the city (see Simmel, 1903/1995; Wirth, 1938, p.197).

For Simmel and Wirth, the city intertwines and amasses people – and, for this reason, there is the need for a greater exactness of co-ordination and organization. The traffic-light is one example of this. From this perspective, it controls flows of people and traffic from one place to another – and traffic-light technologies have become increasingly more sophisticated over time. Moreover, traffic-lights mark intersections in the city where flows come together, where they are intensified and concentrated, where they would interfere and conflict if they were not regulated. The traffic-light shows, it might be added, that it is no accident that people should begin to think of the city as a rat race, simultaneously speeded up, carefully controlled and very crowded. But the idea of density has also entailed another aspect of city life: heterogeneity.

3.3   THE HETEROGENEITY OF CITIES

While describing the social patterns of urban space, Wirth suggested that the city comes 'to resemble a mosaic of social worlds' where the juxtapositions between one piece of the mosaic and another are abrupt and clear-cut. It is important to note, however, the hesitancy in this statement. For sure, Wirth was arguing that urban spaces become differentiated, but he was also arguing that the city should be seen as a place of heterogeneity. This is Wirth's third characteristic of urban life.

Cities only resemble urban mosaics. As we have seen in the case of Chicago's 'Little Sicily' – which looked to outsiders as if it was only inhabited by impoverished Italians – the city is also characterized by the possibility of encounters between people from different backgrounds. Of course, there are many situations in which different people can meet and mix, but Wirth is more interested in the consequences of such encounters. What are these? Wirth argues that the heterogeneity of urban populations causes old rigidities of identity – for example, those formed around class – to begin to break down. It is not that differences disappear, but that they become much more nuanced, more mixed up.

The social stratification of cities becomes increasingly difficult to determine because people are continually straddling and crossing social hierarchies. People's statuses change, from context to context, and over time. As a result, people have to accept that there is an instability and an insecurity in urban identities, formed around any customary hierarchy or social difference. For Wirth, this means that urbanites have a greater sophistication and cosmopolitanism, because they have neither a single allegiance to a particular group nor a single social status.

Because urban identities are exposed as social fictions, individuals are freed to identify, affiliate and associate with divergent groups. The consequences of this are not just personal, however. Urban spaces, like urban social hierarchies, are liable to be fluid, unstable and contain people with allegiances and affiliations to multiple groups. It would, therefore, be a mistake to characterize the city as having a stable pattern of differences – like a mosaic of coloured ceramic tiles cemented into the earth. Instead, the surfaces of the mosaic are liable to shift, to slide over or between one another, even to change pattern. Perhaps this is why Burgess's map of Chicago's social structure so awkwardly overlays two distinctive patterns.

The important conclusion to draw is this. By saying that cities are heterogeneous, Wirth demonstrates that social differences are not the outcome of contrasts between people's innate characteristics or their fixed and unchanging personalities. Instead, social differences are formed through relationships to other people. In the city, heterogeneity tends to break down stable identities because there are great opportunities for people to form relationships with others: to meet, to mix – and to change. These opportunities arise because, in cities, there are large numbers of people, they come from diverse backgrounds and they are concentrated in small areas. Still, we should be cautious about assuming that all cities are like this or that all encounters between different people are equal and have positive outcomes.

Looking back over Wirth's three characteristics of the city, we can see that cities can be something more, through

● bringing together large numbers of people

● the variety of networks through which people intersect (or are kept apart)

● the opening up the possibility that, by bringing differences together, there will be new forms of social interaction, new forms of difference, new opportunities for people to live their lives differently.

However, we should be wary of seeing the city only as a site of volatile social interactions, out of which are produced an inevitable novelty and tolerance. Nor is the city simply a place where all social interactions are indifferent and superficial, so that the possibility that something new or different might be created is closed down. Both these sides of the city have to be kept in mind. Therefore, it is necessary to think about how the intensity of city life is produced while at the same time considering how varied people's experiences of the city can be. In this paradoxical situation, we can glimpse the distinctive character of urbanism. Let us conclude on this point.

4 *Conclusion*

In this chapter, there have been many answers to the question 'what is a city?' We have, nevertheless, been able to identify some aspects of cities which we can take, for our purposes, to be definitive. In particular, we have seen that the city is an assemblage of physical features, the geographic focus of multiple networks and a way of life. Yet, as we considered what a city is, our descriptions became more and more elaborate. Section 1 stressed both the multiplicity of physical features and experiences in cities and also the vibrancy and drama of city life. Meanwhile, section 2 demonstrated that cities are the focus for many networks and that the nature of these networks has consequences for urban development. Understanding cities, therefore, requires a geographical imagination capable of looking both beyond the city and within the city. Further, section 3 showed that there is a manifest intensity to urban life, which is related to other key features of the city, especially their size, density and heterogeneity.

As you look back over this chapter, you will find many more words that have been used to describe cities: such as decadence, vitality, creativity, pace, variety, indifference, and so on. So, are we saying that the city can be defined by lumping all these – and yet more – elements together? Well, yes and no. Yes, because one characteristic of cities is that they contain a range of physical features and a panoply of human experiences. On the other hand, remember, section 1.1 argued that the city should be seen as *combining* physical features and people's experiences *in particular ways*. In order to understand cities, then, it is necessary to appreciate how and why cities combine things (or keep them apart).

In this chapter we have examined how cities bring things together by elaborating on Mumford's idea of the city as a 'geographic plexus'. We have taken his idea further by suggesting that cities are the focus of multiple networks and that these networks need constant maintenance and we have noted that they are always liable to change. From this perspective, understanding cities requires an appreciation of the ways in which networks spread across, intersect with, or avoid, one another (we will pick up this issue in Chapter 3 of this book). It has also been argued that networks both stretch beyond cities, and also intertwine within them. With this in mind, it can be seen that networks also produce differentiated urban spaces – as we have seen in Chicago's case, with its lumber-yards and streets lifted out of the mud, with its Gold Coast and its slums.

Perhaps the most significant effect of bringing things together, of concentrating things in urban space, seems to be intensity. Cities are intense, but this intensity is not a fixed barometer against which to measure whether a place is a city or not. Instead, intensity is the result of what happens when large numbers of people are brought together, in confined spaces; that is, it emerges out of the social interactions within the city – and these might just as easily be mundane and superficial as exciting and dangerous. As we have noted in section 3, living in cities has its consequences for people's way of life. They are brought into close proximity with people who might be very much richer or poorer than they; or from an entirely different country; or have completely opposite views on lifestyle politics, religion and so on. We have noted that the consequences of this are uncertain – it is impossible to predict how any individual will react to all this diversity. But it does suggest that the way the city exaggerates and contrasts, disorganizes and reorganizes, and changes speed (faster or slower), manifests itself as an intensity that cannot be found elsewhere.

Let us return to New York (section 1.1). How does New York figure in your imagination? Does its skyline represent power, dynamism and easy living – the Big Apple. Or is it a Rotten Apple, riddled with unashamed greed, extreme inequality and racial tension? Either way, would you walk in Central Park at night? New York, then, is paradoxical. Both because there are conflicting images of it, and also because people's experiences vary, depending on who they are, what they're doing there, and when and where they are in the city.

So, we can conclude that the city is an assemblage of seemingly contradictory social relationships. More than this, it is through these social relationships that the distinctive character and intensity of city life really comes to the fore. Yet, these relationships are not self-evident. Are cities exciting or dangerous? Perhaps they are exciting because they are dangerous. Perhaps they are dangerous because they are exciting. It would depend, wouldn't it? On who you are. On where you are. On what time it is. And on what you're up to. As we noted at the outset of this chapter, the city unfolds a million different stories. But we have begun to note that the city offers up a series of paradoxes – like that between the individual and the collective, or that between organization and disorganization, or between flows and fixity – that people habitually negotiate, perhaps without even noticing or caring. It is to these mutable experiences of the city that we turn in Chapter 2.

References

Boyer, C. (1983) *Dreaming the Rational City: The Myth of American City Planning*, Cambridge, MA, MIT Press.

Burgess, E.W. (1925/1984) 'The growth of the city: an introduction to a research project' in Park, R.E. *et al.*, pp.47–62.

Castells, M. (1990) *The Informational City: Economic Restructuring and Urban–Regional Development*, Oxford, Basil Blackwell.

Castells, M. (1996) *The Information Age: Economy, Society and Culture*, Volume 1: *The Rise of the Network Society*, Oxford, Basil Blackwell.

Çelik, Z., Favro, D. and Ingersoll, R. (eds) (1994) *Streets: Critical Perspectives on Public Space*, Berkeley, CA, University of California Press.

Cronon, W. (1983) *Changes in the Land: Indians, Colonists, and the Ecology of New England*, New York, Hill and Wang.

Cronon, W. (1991) *Nature's Metropolis: Chicago and the Great West*, New York, W.W. Norton and Co.

Cronon, W. (1994) 'On totalization and turgidity', *Antipode,* vol.26, no.2, pp.166–76.

Davies, W.K.D. and Herbert, D.T. (1993) *Communities within Cities: An Urban Social Geography*, London, Belhaven Press.

de Certeau, M. (1984) *The Practice of Everyday Life*, London, University of California Press.

Grace, H. (1997) 'Icon House: towards a suburban topophilia' in Grace, H., Hage, G., Johnson, L., Langsworth, J. and Symonds, M., *Home/World: Space, Community and Marginality in Sydney's West*, Annandale, Pluto Press, pp.154–95.

Graham, S. and Marvin, S. (1995) *Telecommunications and the City: Electronic Spaces, Urban Places*, London, Routledge.

Hall, P. (1992) *Urban and Regional Planning*, London, Routledge.

Hall, P. (1996) *Cities of Tomorrow: An Intellectual History of Urban Planning and Design in the Twentieth Century* (updated edition), Oxford, Basil Blackwell.

Leyshon, A. (1995) 'Annihilating space? The speed-up of communications' in Allen, J. and Hamnett, C. (eds) *A Shrinking World? Global Unevenness and Inequality*, Oxford, Oxford University Press/The Open University, pp.11–54.

Lynch, K. (1960) *The Image of the City*, Cambridge, MA, MIT Press.

Merchant, C. (1994) 'William Cronon's *Nature's Metropolis*', *Antipode,* vol.26, no.2, pp.135–40.

Mumford, L. (1937) 'What is a city?' in LeGates, R.T. and Stout, F. (eds) (1996) *The City Reader*, London, Routledge, pp.184–9.

Page, B. and Walker, R. (1991) 'From settlement to Fordism: the agro-industrial revolution in the American midwest', *Economic Geography,* vol.67, pp.281–315.

Page, B. and Walker, R. (1994) '*Nature's Metropolis:* the ghost dance of Christaller and Von Thünen', *Antipode,* vol.26, no.2, pp. 152–62.

Park, R.E. (1925/1984) 'The City: suggestions for investigation of human behavior in the urban environment' in Park, R.E. *et al.*, pp. 1–46.

Park, R.E. and Burgess, E.W. with McKenzie, R.D. and Wirth, L. (1925/1984) *The City: Suggestions for Investigation of Human Behavior in the Urban Environment,* Midway Reprint, Chicago, IL, University of Chicago Press.

Robson, B.T. (1975) 'The urban environment', *Geography,* vol.60, pp. 184–8.

Savage, M. and Warde, A. (1993) *Urban Sociology, Capitalism and Modernity,* Basingstoke, Macmillan.

Scranton, P. (1994) 'Commerce and manufacturing in *Nature's Metropolis'*, *Antipode,* vol.26, no.2, pp. 130–4.

Simmel, G. (1903) 'The metropolis and mental life' in Kasinitz, P. (ed.) (1995) *Metropolis: Centre and Symbol of our Times,* Basingstoke, Macmillan, pp.30–45.

Smith, D. (1988) *The Chicago School: A Liberal Critique of Capitalism,* Basingstoke, Macmillan.

Tuan, Y-F. (1978) 'The city: its distance from nature', *The Geographical Review,* vol.68, no.1, pp.1–12.

Wirth, L. (1938) 'Urbanism as a way of life', *American Journal of Sociology,* vol.44, pp.1–24.

Zorbaugh, H. (1929/1983) *The Gold Coast and the Slum: A Sociological Study of Chicago's Near North Side,* Midway Reprint, Chicago, University of Chicago Press.

CHAPTER 2
Worlds within cities

by John Allen

1 *Introduction*

In this chapter we will be taking a closer look at what was, perhaps, one of the most puzzling aspects of trying to answer the question that was posed in the previous chapter: what is a city? The surprise, for the most part, was that the more that was expressed about the city, the more difficult it became to pin it down, either in words or images. The intensity and sharp focus of city life, brought about largely through the proximity and density of social relationships as well as their marked diversity and difference, gave us something to hold onto. Yet much of city life still seemed to slip through our fingers. But perhaps that is just how it is. In truth, it may be easier to acknowledge the fact that we cannot grasp the city as a whole, precisely because it is not a singular entity. There is, then, no one thing called 'the city' which we can simply reveal in all its breathtaking fullness.

Paris, in that sense, is more than the sum of its interactions, and so too is Berlin or Beijing, São Paulo or Sydney, or for that matter virtually any city that you care to mention. They are all places of plurality, a mix of spaces which – as the title of the chapter suggests – represents 'worlds within cities'. These worlds are the central concern of this chapter and the aim is to explore how they are composed and what they add to our understanding of the city.

In taking the many worlds of the city as our focus, however, we should be clear about two things.

First, it is useful to bear in mind that a focus on the particularity of city life does not prevent us from generalizing about it: its varied rhythms, its felt intensities, its seemingly different moods, its concentration of possibilities, and more. All of which, as we shall see, can conjure up the feeling that a city can do things to you or for you; that it can shape your experience of it and suggest all manner of possibilities. When Toni Morrison in her novel, *Jazz*, speaks of Harlem in the 1920s as 'thriftless, warm, scary and full of amiable strangers', a city capable at times of opening itself up to you, co-operating, 'smoothing its sidewalks, correcting its curbstones, offering you melons and green apples on the corner', or on other occasions producing a city sky that 'can go purple and keep an orange heart so the clothes of the people on the streets glow like dance-hall costumes', this is as much a city 'out there', shaping the senses of those within it, as it is one shaped by the many black people drawn to Harlem at that time. It is their worlds, their movements, moods, and backgrounds, which composed it.

A second, related aspect that we need to be clear about before discussing the many worlds within cities is what we mean exactly by 'many worlds'. In one sense, its meaning is fairly obvious where, for instance, it is possible in the span of a few city streets to move from one world to the next, as in the case of Chicago's many different migrant communities in the 1920s referred to in Chapter 1. Such a difference in worlds may be cultural or ethnic in terms of language, style and sounds, or perhaps in terms of the design and architecture of buildings. But the notion of worlds within the city is suggestive of more than that, as Toni Morrison's writing conveys and indeed the previous chapter signalled. In addition to thinking about the settled nature of communities and their diversity, therefore, the intention here is to consider the many worlds as fluid in form and cross-cutting in style. Perhaps a useful way to imagine the city in this respect, then, is

- as a series of overlays, where some movements and relationships come into full view whilst others are partially hidden or obscured;

- yet, as the overlays shift, different groups of people come into proximity and different kinds of worlds meet or routinely glance by one another.

Much of this chapter will be spent exploring this representation of city life, in part because it allows us to consider how different groups mix and cross one another in the city, and in part because it draws attention to the lines of inequality drawn through the city. In the next section, we examine the sets of rhythms, the regular flows and patterns, which divide up city life in seemingly routine and often ordinary ways. The movement of groups in and out of focus at different times of the day, and the very real sense of worlds moving through the city, are the central concerns here. Following that, in section 3, we turn to the more settled arrangements of cities, in particular, to the mix of buildings and streets that give meaning to city environments through their symbolic contrasts and juxtaposed styles. The symbolism of the built environment, both expressive and representational, is explored in relation to issues of inclusion and exclusion. Finally, in section 4, we address the issue of how the often bewildering diversity of city life, in all its material and social variety, is itself negotiated. In this way, we examine the processes by which groups deal with difference in the city when their worlds actually meet and the intensities that such processes may arouse.

It is to the substantial nature of city rhythms, however, that we turn first.

2 City rhythms: the comings and goings of city life

By city rhythms, we mean anything from the regular comings and goings of people about the city to the vast range of repetitive activities, sounds and even smells that punctuate life in the city and which give many of those who live and work there a sense of time and location. This sense has nothing to do with any overall orchestration of effort or any mass co-ordination of routines across a city. Rather, it arises out of the teeming mix of city life as people move in and around the city at different times of the day or night, in what appears to be a constant renewal process week in, week out, season after season. Jane Jacobs in *The Death and Life of Great American Cities*, observing street life in New York in the 1950s, likened the ebb and flow to an 'intricate ballet' in which individual dancers with their own choreographed parts moved around and across one another to compose a daily dance of the street. Through an intricate pattern of movement and change, improvization and ritual, a mix of people doing little more than going about their daily business was seen to animate city life. For Jacobs, the intersection of the different rhythms broke up the day into a variety of street ballets, each scripted to make the whole ensemble meaningful in an urban way. She confined her observations to street level but, broadened out to the city at large, such a mix of rhythms can be seen to capture many of the flows which mark city life whilst, somewhat paradoxically, only revealing them in part. Consider, for instance, the following rhythmic landscape.

2.1 A DAY IN THE LIFE …

Imagine a city at five in the morning. There is a certain stillness in the air, punctuated only by the dull sound of the occasional truck or car. In the half light, it is possible to make out the profile of a bus as it winds its way towards you from the suburbs, stopping briefly here and there to set down its load onto the quiet city streets. At this time of the day, it is mostly women and young men who descend from the bus. Many of them are cleaners, kitchen-hands and other less glamorous service-workers who make up part of that nebulous workforce that brings a city's empty buildings outwardly to life each day. Snatches of conversation on the street with familiar others, the odd shout of recognition, can still be heard above the stop–start motion of the traffic and the daily clatter of delivery vans as they go about their meandering business. It is not yet rush hour and, on the face of it, the city is clearly not that rushed.

If there were hills in view, a clear skyline maybe, this sketch could possibly be Lisbon or even Lyon, but without too much imagination it could also be parts of Manchester or Manila. In fact, it could almost be any place where the

FIGURE 2.1   *Early morning street scene in Rome*

intermittent sound and slow tempo of street movements, not to mention the many dawn smells, mark out this time of the day in the city for particular groups of people. Of course, individual cities combine different kinds of movements to evoke quite distinct senses of rhythm. There is no one overbearing city beat. And yet, there *is* something broadly recognizable about how such rhythms fall into distinct times of the day. Mix the transport modes, for example by adding boats and ferries (and the wail of their horns) to the staple thoroughfare of road traffic and then the early morning rhythms depicted would not be out of place in Hong Kong, with its population of island-hopping inhabitants. Alternatively, take away the water transport and substitute them with swarms of bicycles, plus lines of motor vehicles which display the ubiquitous VW logo, and the early morning rhythm would be easily recognizable in Shanghai.

By eight or nine in the morning, however, many cities take on a rather different feel and a different set of associations. Among the more obvious is that the noise-level rises, as different kinds of sounds, many of them generated by the increased flow of traffic and people, distort and blend into one another. The result, in a great many cases, is an almost indistinguishable cacophony of sounds. Some of the sounds are all too familiar in many cities, such as the pulsing surge of cars accelerating or the pounding vibration of overloaded lorries. Other sounds are less distinct (was that a car door slamming? or a load being dropped on the roadside?), and can only be guessed at. In places like Bombay (Mumbai) or Calcutta, we can add the ever-present sound of car horns honking, radios blaring and the livestock which add their own distinctive contribution. Alongside the rise and fall in the intensity of noise, however, is the wearily predictable smell of exhaust fumes, much of it a diesel vapour, that assaults the nose.

FIGURE 2.2 *Calcutta street scene*

Move away from the deafening streets of Mumbai or many a European city to the freeways of a North American city, however, and any potential assault on the senses is likely to be lost on car-drivers who, in Richard Sennett's terms, want to pass through city space, rather than be aroused by it. As he observes in *Flesh and Stone*:

> The physical condition of the travelling body reinforces this sense of disconnection from space. Sheer velocity makes it hard to focus one's attention on the passing scene. Complementing the sheath of speed, the actions needed to drive a car, the slight touch on the gas pedal and the brake, the flicking of the eyes to and from the rear-view mirror, are micronotions compared to the arduous physical movements involved in driving a horse-drawn coach.
>
> (Sennett, 1994, p.18)

But if those who drive vehicles (especially air-conditioned vehicles) are largely removed from the passing scene, the same can hardly be said for those walking the streets in the morning rush. A whole range of stimuli is potentially there to confront and envelop pedestrians, from the scurrying to-and-fro of office-workers in the commercial districts of cities, to the bustle of the streets themselves, as tourists, shoppers and street vendors jostle for position in the pursuit of their single-minded goals, often cutting across the lines of one another in the process. On the crowded walkways, bodies may touch or glance past one another in the most personal of ways, yet routinely elicit little or no response. A glimmer of recognition here, an affectation of indifference there, but little that could possibly pass for involvement. And, of course, in the background, is the constant cacophony of sounds. In such potentially overwhelming situations, it is all too easy to shut out the noise after a while and to lose touch with what is happening directly in front of you.

At this time of the day especially, the presence of all different kinds of people going about their daily lives may fail to register. The homeless wandering the streets in many a downtown area, the armies of night security guards leaving their so-called 'empty' buildings, or the early-morning cleaners making their way home after having polished the city's offices – all may go unnoticed at this time of the day, swallowed up by the rhythms of those going in other directions.

No two cities are alike in the precise combination of rhythms and movements which mark out this shift from the early morning to morning proper. And yet, in every city it is possible to trace the presence of different groups moving in and out of focus as the day progresses. In the heart of São Paulo for instance, it is the unofficial street-traders who fade from view as the formal working-day gathers momentum. At its centre, in the Anhangabaú Valley, Elisabetta Andreoli in the appropriately titled text, *Strangely Familiar*, has charted the comings and goings of those who make up the different times of this city space:

> In the early hours, hundreds trade goods – imported, smuggled or just obtained without the necessary documentation – on the pavement in front of closed shops. Improvised and short-lived kiosks sell hot drinks and cakes. When the shops open, these early traders yield to others selling the same products but with official licences to trade. During the week, hoards of bank employees, businessmen and administrative staff populate the area while street-children try to get their bit out of passers-by. At the weekends, Nordestinos (immigrants from the north) turn the less noisy space into their own territory for specific trade or religious gatherings. Sharing, or rather competing, for the same space generates tensions and in the street police and even private guards make their presence felt.
>
> (Andreoli, 1995, p.66)

FIGURE 2.3 *Street traders in early morning São Paulo*

In this snapshot, each part of the day, and indeed each part of the week, gives
way to the next as groups displace one another or compete for the same space.
Early morning vendors trading informally give way to their formal namesakes
and businessmen who claim the space as their own, yet are forced to share it
with street-children, and so on. With the onset of the formal working-day, the
place takes its rhythms from the five-day working-week, giving the impression
that the business and professional classes have it all to themselves. In living the
illusion, they attempt to erase the traces of others, such as the street-children,
even by force of the law if necessary. Whether successful or not, the movement
of some groups in and others out of focus in this way, does not mean that many
simply disappear from the city or are somehow rendered invisible. Rather, they
move unseen and unnoticed. They are 'out-of-place', so to speak, even though
they may actually live and labour there.

For much of the day, in busy, sprawling cities such as São Paulo, it is difficult to disentangle the many different rhythms and movements by which groups move across one another. Whilst rhythms may be thought about in a general sense as composing city life, their timing varies as much within cities (for example, between dense city centres and outlying suburban reaches) as it does between them (contrast Spanish cities with their northern European equivalent in the mid-afternoon, for instance). By nightfall however, as dusk begins to settle, many an observer has drawn attention to the city as a different place to be.

The most perceptible shifts in mood are perhaps directly attributable to the change in light. The neon skyline, the fluorescent-lit office-blocks which give the cityscapes of Hong Kong, Jakarta or New York their distinctive feel, is one such expression. Another is the darkness of the empty squares or deserted shopping-centres late at night which can turn the experience of walking into an anxious, rather than a sociable experience. In empty, poorly lit locations across many a city, in the spaces of daytime leisure for example, the stillness of the late hours may act as a form of cover for, at best, harmless excitement or, at worst, violent crime. In the former landscape, the intensity of illumination – the powerful streetlights, the gleaming advertisements, the brightly lit shops – can transform not only the lines of buildings at eye-level, but also how people behave. In such a landscape, there appears to be a different tempo to the place, almost a buzz, a sense of vibrancy that distinguishes the crowds here from those further out. Nightlife in New York, for instance, with its endless entertainment opportunities and 24-hour service draws attention to its noisy revellers and to that hum of excitement that is inextricably linked to the centre of such cities. But it also draws attention to those working in the all-night grocery stores, and the many clubs, restaurants and service-stations. Whereas for one group the night is likely to curb many of their urban pleasures by late evening or early morning, for the other, the night stretches their working routines through till dawn in a monotonous sequence.

Murray Melbin in *Night as Frontier* talks about the different groups which come to the fore after dark in this way as a kind of 'colonization' of the night. For Melbin, the spread of activities through the night gives rise to a series of displacements as groups move across one another in time and space:

> Chains of events link up according to the logic of their purpose. Companies specializing in cleaning offices send porters, cleaners and maintenance men out after business hours. Day and night progressively adapt to each other and enter a fuller partnership. The endeavours themselves are not continuous. They join in appropriate little sequences that displace them into later or earlier times. Some connect haphazardly, as when a late-phased and an early-phased activity reach so far in their own directions that they overlap. They have spread out until they result in overnight connections between one day and the next.

Late-phased events trail the profusion of daylight affairs. Since most people work during the day, their recreation is shunted into the evening. As evening arrives the theatres, cinemas, sports arenas and neighbourhood bars begin to thrive. So do the cocktail lounges, cabarets, bordellos, and private drinking clubs. Few of these activities last all night. They take their turns, and as they end other projects begin. Restaurants, dance halls, and public bars shut their doors in the early morning. Private sanitation companies wait for them to close before beginning to collect trash, so that customers will not be disturbed. In San Francisco trash trucks being collecting at 4 am. In Paris street cleaners with wheelbarrows and long handled shovels are on their rounds by five in the morning.

(Melbin, 1987, pp.82–3)

Rather than the different activities running together and merging, however, what is more important to stress is the haphazard nature of the relations and their overlap. Even in Jakarta where wealth and poverty are mirrored in the city's nightlife – with roadside entertainment and cheap 'dangdut' bars on the one side, and upmarket, western-style hotel nightclubs on the other – there is little that brings these activities together. While a number of the activities identified by Melbin do follow on, one from the other, through the night, many of those involved – in transport, trading or selling, for example – are in fact moving to a separate but related set of rhythms from those in the theatres and bars. If they should meet, then the overlap is more likely to be similar to that of the crowded walkways, where close proximity elicits little in the way of a response. In that sense, it is not just the cover of darkness which appears to conceal movement in the city at night, it is also the inability of different groups to acknowledge the presence of others.

Much the same was noted about events in central São Paulo in the early morning, where indifference not invisibility was more the issue. And this indifference can be seen to translate into cultural and economic terms at night where, the later it is, the less social the hours, the more likely that those working are disproportionately drawn from the ranks of the poor, the insecure and the recently migrated. This is not simply to affirm that Koreans run the all-night grocery stores in New York or that Algerians swell the small army of street-cleaners in Paris in the early hours, for there is no general rule in play here. Rather the point is that the daily rhythms and movements of cities routinely code and divide city space on an unequal basis that is rarely acknowledged.

From this sketch of city rhythms, therefore, it is possible to see:

- how the worlds of different groups in the city may routinely overlap yet remain apart from one another

- that cities bring all kinds of relationships into close proximity, yet, by virtue of the juxtapositions produced, create the possibility for indifference and exclusion as much as they do Jane Jacobs' meaningful street-scene.

There is a lot to consider here, so let us pause and take a closer look at what rhythms encompass.

2.2   SENSES OF RHYTHM

One way of thinking about city life is thus in terms of the regular beats and repetitive flows which mould it in different ways at different times. City relationships are far from flat or monotone in their sequences. The constant succession of movements, their changing pace and tempo, can tell us something about the broad 'feel' of individual cities at different times of the day. This broad 'feel', as we have seen, is made up of an overlay of relationships. But, as also indicated, there is more to rhythms than simply a patterned outcome.

It is probably useful, therefore, to start with relationships that lie at the heart of city life.

In the *first* place, the rhythmic movement of cities provides us with an insight into the ways that people may encounter one another in the city. Many are unfocused, for instance, although perhaps no less intense because of that. Sitting on a bus or a train or tram for a matter of minutes or hours, for example, without talking to those around you, entails certain social consequences. As indeed does moving among others along crowded streets and bustling walkways in a detached manner. One such consequence, according to Georg Simmel (1908a/1969), a turn-of-the-century urban social theorist, is the ever-growing attachment to visual interaction. The quick nod, the measured glance, the look down, or the mutual eye contact which loses its directness at that very same moment, are all taken by Simmel as an indication of the remoteness of city life. For him, the city is more a world of glances, of visual impressions, where people distance themselves from others in forced proximity in order simply to get by. A degree of social reserve, therefore, on this account, is a necessary part of what it means to live in a city.

The reverse may also hold however, as was indicated in Jane Jacobs' account of New York's 'intricate street ballet'. In contrast to Simmel's projection of detachment, she showed how the daily ebb and flow mixed up old and young, professionals and shopkeepers, locals and passers-by, in a meaningful street-scene which owed as much to chance encounters and visual cues, as it did to changes in the pitch and tone of the street. People co-exist, not as intimate associates, but as relative strangers, often exchanging meaningful eye contact, with the locals especially listening to the street, gauging its different moods and dispositions.

Clearly, the specific mix of rhythms in different parts of cities can give rise to a diverse range of encounters. The sequence of displaced relationships referred to by Melbin (1987), for example, is one such remote form and Jane Jacobs' meeting-points simply another. What they all draw attention to, however, is the importance of the senses to an appreciation of city rhythms.

63

The *second* point to note, then, despite Simmel's worry that we may be reduced to seeing rather than hearing people in the city, is that all the senses come into play in relation to understanding city life. Henri Lefebvre (1992, 1996), in one of his last works, expressed this forcefully by drawing attention to the fact that the experience of the city is more than what we are able to see. It is also, as Jane Jacobs intimates, about what distracts or assures us through its familiar and not so familiar sounds and smells. On listening to the city:

> If one attentively observes a crowd during peak times and especially if one listens to its rumour, one discerns flows in the apparent disorder and an order which is signalled by rhythms: chance or predetermined encounters, hurried carryings or nonchalant meanderings of people going home to withdraw from the outside, or leaving their homes to make contact with the outside, business people and vacant people – so many elements which make up a polyrhythmy. The rhythmanalyst thus knows how to listen to a place, a market, an avenue.

(Lefebvre, 1996, p.230)

Listening to a place, for Lefebvre, is thus a particular state of awareness; as he suggests, it is more about 'moods than of images, of the atmosphere than of particular spectacles' (1996, p.229). This active, expressive dimension comes close to the sensual dimensions of the city highlighted earlier in the description from Toni Morrison's novel, *Jazz*, and can include the sweet and not-so-sweet smells of, say, an early morning market-place, as well as any changes in the dawn light or resonance to the ear. Although not on the same sensual level, Lefebvre believes that it is equally possible to 'listen' to a street or a location in a city by tracing the rhythms expressed in a particular mood – their movements up and down, and their smooth or jerky tempo.

ACTIVITY 2.1   Think back to the previous section on the sequence and mix of rhythms which can characterize city life in the course of a single day. Take a city that you know reasonably well and consider whether or not it has a particular mood or atmosphere – maybe at a particular time of the day, in a particular part of that city. How would you express that feeling? How would you convey to others that, looking at a postcard, say, or a photograph of that particular place, was not quite enough to capture all that is involved in experiencing such a place? ◆

It is a difficult point to put across but perhaps the key issue to stress is that, in attempting to understand city life, it is important to recognize that what makes the city is more than a string of visual images, representations and encounters. As indeed the previous chapter emphasized, cities are sites of embodied intensity: that is, they are heard as well as seen, and taken in through the nose as well as the eyes.

This leads to a *third*, general point which was evident throughout the earlier account of the comings and goings of those who live and work side-by-side within cities: namely, the question of who is seen and who is heard in the daily life of cities. Lines of visibility and waves of sound are also forms of inequality, whether in São Paulo or Shanghai, Paris or New York. The issue of whose rhythms appear to come to the fore is directly related to that of power and the ability of certain groups to superimpose their rhythms on others. Again, the work of Henri Lefebvre is useful. In *The Production of Social Space* (1991), for instance, he sets out to show how spaces are coded by dominant rhythms which are able to give the impression that in a particular space, or in a particular building even, only certain groups are actually present. The centre of many a city, for example, has its commercial districts which signify what kind of activity is appropriate where and, as before, who is 'out of place'. It is this ability to *smother* difference, to suggest who should be seen and heard and who should not, that can give parts of cities the impression of sameness rather than displacement and diversity.

At its most obvious, we have seen this in the case of those who produce the city through their unseen service work, in the twilight hours in the neon-lit offices, for example, or in the back-end kitchens of hotels and restaurants. But it is no less evident in the case of São Paulo's street-children or indeed in the many informal activities which characterize most cities. To run ahead of ourselves, perhaps one of the most revealing examples of late, in cities as far apart as Istanbul and Singapore, is the number of Filipina women working as domestic labour whose presence is barely registered, let alone seen. Working as live-in maids and housekeepers, they move unnoticed among themselves all week, only to erupt on the public scene on Sundays (their one day off) by spilling noisily onto the streets outside Catholic churches in numbers that defy any dominant coding of space. Some 16,000 Filipina migrants in Singapore, for instance, bring a sense of rhythm that is altogether different from that superimposed by the movements of the many tourists, shoppers and traders going about their routine business.

In thinking about how rhythms express or represent a city, therefore, we need to be aware:

- that it is usually only certain movements and relationships that come into full view
- that it is a selective process as to which aspects of the different worlds are raised to our senses, and
- that it is difficult to grasp the city as a whole because much of what composes it is routinely obscured or unacknowledged.

2.3 WHOSE RHYTHMS?

In part, this is the question that we have been addressing all along. That the rhythms and movements of cities routinely divide up city space on an unequal basis has been stressed. That there is some kind of shifting overlay of elements which brings relationships into proximity – sometimes as if without touching, at other times to displace and modify – has been illustrated. But what gives particular rhythms or sets of rhythms their impetus or dynamic? And, more importantly for our concerns here, what kind of city spaces are we talking about when they cross over one another in the ways that have been suggested?

You may have noticed in the account so far that it has become increasingly difficult to view the city as we would a map: as a flat, unbroken surface, with a broad sense of how the different parts of a city, its various communities and districts, relate geographically to one another. Whatever you may think about maps, the kind of overlapping rhythms and movements that we have considered have tended to disrupt the unbroken surface of cities and indeed, bring into focus the fluid and cross-cutting nature of city life. Here, we want to push this line of thought further by looking at how city spaces are constructed in a discontinuous fashion. Put another way,

- city spaces are not simply given: they are produced through many movements and interactions coming together in ways that often *disrupt* existing rhythms and relationships across cities.

If we return to the example of Filipina migrants working as housekeepers or nurses abroad, we can begin to grasp what is involved.

Perhaps the first thing to note, as Petra Weyland (1997) points out, is that being a global housekeeper or nurse – in Istanbul or Singapore, or indeed in Rome, Madrid, Tokyo or Jedda – is a more worthwhile economical activity for Filipinas than is working in a global factory back home, and it affords them greater social status too. On that basis, the large transnational flow of Filipina domestics to affluent, mainly western and middle-eastern cities, leaves its mark on both the cities they come from and those to which they migrate on a temporary basis. For despite the lack of public presence, as indicated above, the Filipina migrants add a new dimension to the private, domestic spaces of households, certainly in places like Istanbul. As Weyland goes on to argue, the symbolic capital of Istanbul's global business community has been re-scripted by the arrival of Filipinas in their private households, as other 'local' live-in help is displaced by the status attached to their 'around the-clock' services. And this is no less significant because of the way in which they have been made less visible by Istanbul's gendered public spaces, where both they and local women do their best not to be noticed.

If it were, say, the impact of the Islamic diaspora under scrutiny, the situation might well be reversed. Consider, for example, the highly visible disruption brought about by the construction of a minaret in a previously non-Muslim area of a city. The introduction of a different set of public rhythms, around the call to prayer, for instance, and its particular geography, stands in sharp contrast to the possibility that many of those attending the mosque may be from among the 'hidden' service workforce working 'behind the scenes' at airports, hotels and the like. Commenting upon the impact that such a public construction as a minaret can have in a western city, Gilsenan draws a direct analogy from that of the colonial city:

> Imagine – and it is very difficult for those who have not experienced the world of the colonized – the effect that outside forces, over a relatively short period of time, can have on the transformation of the *whole* of the relations that make up urban space, including its social geography and unquestioned givens of the way things are in cities. Imagine, not only one building being constructed on an alien model, but an entire system of urban life in its economic, political, and symbolic – cultural forms being imposed upon already existing towns and cities that have been organized on a different basis.

> (Gilsenan, 1982, p.195)

Imagine, then, the daily rhythms of a city as in many ways interrupted by relationships and codes drawn from elsewhere, from outside of the 'familiar' so to speak. Such examples, where the very sense of time and space across a city as a whole may be interrupted or overturned are extremes of course. But they are no less common for that.

ACTIVITY 2.2   You can judge this for yourself by reading Extract 2.1 by Ulrich Mai on culture shock and identity crisis in East German cities in the aftermath of the fall of the Berlin Wall in 1989. As you do so, make a note of the changes in the relationships of daily life and the disruptions to the familiar rhythms of city life involved. ◆

EXTRACT 2.1
Ulrich Mai: 'Reconstructing the urban landscape: culture shock in East German cities'

Since the political changes in 1989 the reconstruction of East German cities has been carried out with breath taking efficiency and speed, leaving the ubiquitous and obtrusive traces of the new economic system … Buildings under repair disappear behind scaffolding to receive new paint, roofs and windows, not to speak of the many new buildings which have already been constructed: hotels, department stores, videothèques and so on. Interestingly enough, immediately after unification considerable attempts were made to support the enjoyment of car-ownership, which after the long years of scarcity was widely considered a central symbol of a higher standard of living. There was not only an immense invasion of car traders, garages and service stations, but also complementary swift action was taken to replace the old bumpy cobblestones with new tar pavements, and even the government was eager to replace almost overnight the old traffic signs made of plastic with solid, Western-style ones with a new colour and design, while on the outskirts and in the countryside many an attractive side-street was sacrificed for the sake of wider roads in order to carry more traffic.

Most obtrusive, however, are the changes of the urban landscape in the central business districts (CBDs). While in socialist times CBDs used to be hardly more lively or colourful than any other part of the city, within a few years they have come to abound with all the signs and symbols of the capitalist commodity worlds and its multiplicity of colourful and promising gimmicks. Nowhere else has the change in the environment been more drastic or faster. No wonder local residents often express difficulties in spatial orientation, in 'finding their way'. In fact, until recently, the East German city, and especially the historic city centre, was unique in the sense of bearing its unmistakable individual face. Swift modernization with its equalizing universal standards that are indifferent towards spatial idiosyncrasies has, however, destroyed uniqueness and imposed uniformity. Within only four years, cities in East Germany have come to resemble their universal Western model.

Strange symbols, alienation and symbolic expropriation

No doubt, the reconstruction of the East German home meets with the wide cognitive approval of the population, as it reflects above all the overthrow of an unpopular political system. This does not, however, exclude a deep identity crisis and feelings of suffering elementary losses, even feelings of alienation and expropriation.

The issue seems most plausible in the depletion of the old social environment. First of all there has been a remarkable deterioration in the social infrastructure: kindergartens were closed down for reasons of 'economic effectivity', and the same happened to youth clubs, recreation centres, even pubs, small retail shops and post offices which all also used to serve a social and communal purpose ... In a characteristic case an old community hall, where virtually all private and public festivities used to take place, was turned into a more profitable furniture depot, leaving the community with no place to hold celebrations.

Of course, unemployment, an experience hitherto virtually unknown, represents the most drastic impact on local identity, as it cuts off familiar social links that used to go far beyond mere work relations to strengthen group solidarity and integration in locally-based interaction …

Also, there is a noticeable change in the construction and composition of personal networks. After the political overthrow their major purpose – to pool and transact resources in order to acquire scarce goods – has become obsolete in the market economy. Instead there is a continuous need for information job opportunities, cheap credit and, of course, for reliable assistance on how to learn many modern skills, such as how to fill in the unavoidable tax-return forms and how to identify the better of two competing offers from different health insurance companies. Evidently, as the type of social capital in social relations changes, personal networks spread in terms of competence represented by members and in terms of the space over which members in networks are connected. Thus, as networks are restructured under the new political and economic conditions, the old quality of locality becomes less important. Only with the losers in unification – the jobless and the old – do the locally focused personal networks, spatially confined to neighbours, relatives and close friends, survive, mainly for the purpose of emotional and to some degree manual support. Generally speaking, however, personal networks in East Germany now reflect the erosion of local identity and, in fact, contribute to its crisis.

Strange symbols have invaded virtually all spheres of life in East Germany so fast and overwhelmingly that people often feel themselves strangers at home … There is, of course, not only the metamorphosis of the urban landscape, the many new façades, commercials, and traffic signs; social and political changes are accompanied by the shock of unfamiliar tastes (food) and even smells (especially of the cleansing agent applied in public places), thus contributing to the loss of sensual orientation or at least causing unknown problems of sensual adjustment in spatial orientation.

What really aggravate the difficulty of readjustment are the invisible symbols and signs of strangeness, such as the innumerable new laws and regulations necessary to register a private car, to receive the rent or tax refund you are entitled to, to claim on an insurance policy (after the many cases of fraud) and so on. Almost overnight there are entirely new bureaucratic rules and institutions and running them, usually in superior positions, are 'experts' from the West, representatives of the new system, whose consciousness of their superior experience and competence often makes them appear arrogant and presumptuous…

The victory of 'the other' political system is, in a subtle way, reflected in the many symbolic gestures of triumph which leave no doubt which side is the winner after the East–West confrontation and the long competition between societal and political systems. There are indeed striking examples: in an act of unreflecting and undiscussed retaliation, conservative majorities in the city parliaments swept away many old street- and place-names that carried a socialist meaning, making no distinction between local communist heroes, Karl Marx or anti-fascists. I do not here question the necessity for a critical moral dispute about history or the role of individuals in it, provided it *is* a dispute. In any case, however, one cannot totally ignore the psychological consequences in terms of alienation; in the process of socialization the individual learns symbolically to 'appropriate' the environment by being able to *name* it, so that it becomes a part of his or her identity. Consequently, the strong feeling of being 'at home' is undermined when the old familiar place-names become obsolete.

But, of course, there are other symbols of political overthrow, many of which attract more collective attention. A case in point is the re-construction of the Garnisonskirche (Garrison Church) in Potsdam which had been pulled down by the communists directly after the Second World War as a pre-war symbol of the historic coalition between the Nazis and the bourgeois conservative forces in the country. Of similar symbolic value are the reconstruction of the baroque Dresden Frauenkirche, the ruins of which many would rather preserve as a war monument, and, above all, the plans to remove the socialist Palace of the Republic in Berlin, a popular multi-functional parliamentary building during the socialist period, and to reconstruct in its place the old city palace of the former Prussian dynasty.

Source: Mai, 1997, pp.76–9

Clearly, the pattern of reconstruction and the disruption to the rhythm and relationships of East German cities is not something that happens to cities every day. But in witnessing the range and scale of disruption to their everyday spaces, and the profound sense of disorientation engendered, even down to the unfamiliar smell of public places, the population of East German cities experienced in a more rapid, intense fashion what happens when one set of rhythms clashes with another – and loses. More to the point, they experienced the pull of a different system of relationships and symbols whose rhythms were effectively governed by an unaccustomed pace and drive.

Take, for instance, the simple interruptions at street-level referred to, not only the removal of familiar place-names, but also the introduction of new-style traffic-lights and wider, tarred throughways – ordinary features of western developed cities – which translated into a faster pace of life and a greater intensity of noise levels. To this we could add the visible disorientation of city-centre renewal as a commercialization of both property and goods took hold, placing the superfluous and the necessary side-by-side in an unfamiliar mix, or we could mention the less visible regulations and bureaucratic ways of doing things which made East Germans feel strangers in their 'own' city.

Perhaps one of the most far-reaching disruptions brought about by the collapse of communism in East Germany, however, concerns the transformation of networks, at city-level and beyond. As a new type of social and cultural capital, drawn largely from the western commercial markets, displaced the predominantly political character of personal networks, so different groups found themselves either in harmony with the new external economic rhythms or by-passed by them. The culture shock in this instance comes about, according to Mai, not merely through the devaluation of certain forms of information, trust and connections, but also through the different backbeat to which successful networks are expected to conform.

This list is far from exhaustive and no doubt you were able to pull out further examples of the dislocation in both space and time which have recently transformed East German cities, but a sufficient number have been raised to give an idea of how the spaces of such cities have been cut across from the outside. It is not only from the outside, however, that cities can experience disruption. It can also come from worlds that have been described as being 'below the surface'.

This may seem a little odd, but think about it for a moment. The possibility of the surface of cities being interrupted by the movements of those who live and work 'below it' – on the subsurface of the city so to speak – comes close to the idea that much of what goes on in cities remains largely out of focus. Or certainly to those who construct themselves as at the centre of cities. How, for instance, do we portray the informal activities which compose much of city life in places such as Mexico City, Bogotá, Nairobi, Jakarta or, less obviously, in Miami or New York? Well, if the example of disruption in East German cities represents an extreme, the kind of interruptions which flow from the informal economy, in, say, Mexico City is best described as *routine*.

The boundary-line between the formal and informal side of cities is far from rigid and is useful only in so far as it draws attention to the unrecorded and unregulated nature of routine economic activities within cities. It is the unrecorded – and sometimes illegal – nature of informal work and exchange which gives it its clandestine or 'below the surface' feel. This can amount to anything from street-trading, for example the many food-sellers who push their mobile kitchens in and out of the passing traffic, to shoe-shining, watch repairs, portering, fetching and carrying, passenger-ferrying, scavenging, as well as prostitution. But it is not only on the streets that the informal side to cities 'reveals' itself, especially in the fast-growing cities of the developing world. The physical presence of shanty towns, whole networks of exchange in themselves, is a reminder that the worlds of formal and informal activity do more than co-exist.

71

In 'probing' the surface of Nairobi, for example, in Mathare, one of its larger shanty towns, Nici Nelson (1997) notes the many activities which brought the formal and informal worlds into contact. Whilst women in the shanty towns are primarily involved in brewing beer, selling vegetables or prostitution, men are more likely to move through the city as construction workers, electricians, glaziers, plumbers, drivers, delivery-men, street-sweepers and the like. The work for the women and men was equally precarious, often a stop-gap measure until something better came along, but no less an integral part of what makes Nairobi the city that it is. Larissa Lomnitz draws a similar conclusion about Mexico City from her observations on a shanty town in its southern half. As she sums it up:

> Most settlers of Cerrada del Cóndor are rural migrants, or the offspring of migrant parents. Most of them are unskilled workers, such as construction workers, who are hired and fired on a daily basis; journeymen and artisans who are hired for specific jobs and have no fixed income; petty traders; and people who work in menial services. They may be described as urban hunters and gatherers, who live in the interstices of the urban economy, where they maintain an undervalued but nevertheless well-defined role. They are both a product of underdevelopment and its wards.
>
> (Lomnitz, 1997, p.216)

FIGURE 2.4 *Mathare Valley, Nairobi*

This, then, is more than simply a tale of rural migrants, newcomers to the city, getting by as part of an urban survival strategy. The demand for much of this unskilled work comes from the very structure of these cities: its movement is built in to the way people, goods and things circulate. And it is the same for powerful cities like New York. As Saskia Sassen (1989) argues, the 'underground' economy in New York City, whilst largely concentrated in migrant communities, is primarily a response to opportunities created by the restructuring of the city's manufacturing economy and the proliferation of small firms in a range of traditional and less traditional sectors. The growing incidence of homeworking arrangements and the increase in 'putting out' work to sweatshops on an informal basis both represent part of the less obvious spaces which make up New York's city economy.

The use of the term 'underground' here to refer to one aspect of the city's economy which is often overlooked is, of course, not a literal reference to a subterranean world (a point incidentally which does hold direct relevance for those living in the railway tunnels beneath Grand Central and other stations in New York: see Toth, 1993). But, like the earlier examples of how the spaces of a city may be disrupted by the coming together of different rhythms and relations, the representation serves to problematize our thinking about how cities and their plural worlds connect. A further reason why much of the life of cities often appears to slip through our fingers when we attempt to grasp it, therefore, is simply because

● their daily rhythms are subject to the kinds of external disruption and routine interruption which stem from the familiar, as well as the less familiar, networks of which they are a part.

3 City environments: juxtapositions of feeling

So far, we have tended to focus on aspects of the city that stress the movement and flow of relationships through them. To that end, we have been thinking about how cities are made and re-made in response to what moves across or disrupts them. But cities are also known for being settled environments with material characteristics that give meaning to those surfaces and tangibility to their expression. Indeed, perhaps more than any other feature of cities, the built environment has the potential to stabilize their image as a whole. Certain buildings or structures often appear to lock a series of traits – architectural, symbolic, material – into a fixed image so that, for instance, the Eiffel Tower is Paris, the Opera House is Sydney, the Petronas Towers are Kuala Lumpur, and so on. Such buildings and structures are best thought of as expressive of the cities of which they are a part; they signify something about what it is to be in those cities, just as the sea of monumental cranes bobbing on the Shanghai skyline signifies all that is currently dynamic and thrusting about this Asian city and its growth. But such images are also – and perhaps necessarily so – a form of shorthand. Few aspects of the built environment can carry this kind of symbolic burden. Such features are the exception rather than the rule. Broadly speaking, cities possess a diversity of building styles and manufactured environments, some monumental in form, others apparently less so, but all more or less transformed through occupation and use over time, and all in their own particular way *expressive of meaning*. The interpretation of which takes us back to the many worlds which comprise cities and to questions of power and proximity.

3.1 BUILDINGS, MONUMENTS AND 'WILD' PLACES

Back in the 1930s, when describing the buildings and streets of São Paulo, Claude Lévi-Strauss, the French anthropologist, spoke about the city as a 'wild' place: not ugly or out of control, but a place full of exaggerations and surreal contrasts in its material make-up and layout. The mix of buildings from different times, with different histories and architectural styles, the juxtaposition of affluent and poor streets, the incongruity of tall concrete blocks next to bamboo-frame shanty housing, all provided Lévi-Strauss with a sense of evocative contrasts and sharp proximities. English-style parks with ornate lawns and statues were to be found alongside the local Automobile Club and the offices of the Canadian company responsible for the lighting system and public transport; avenues of palatial housing would stop abruptly and give way to hillside shanty towns on either side of muddy torrents; local workshops exercised their traditional crafts alongside the trading of Syrian bazaars, both a stone's throw from the only skyscraper in São

Paulo at that time. If this represented an untamed environment to Lévi-Strauss, however, this had less to do with its appearance as an unruly mess and rather more to do with what history had brought together.

In that sense, the odd juxtapositions revealed by the built environment of a city also reveal its different histories. Buildings in particular have the ability to carry the traces of past interactions and how people with different cultures and memories have faced one another in the same city, if not across the same street. Reflecting on contemporary São Paulo, for instance, Andreoli follows Lévi-Strauss in drawing attention to the different times embedded in the buildings at its centre:

> To foreign eyes, the Anhangabaú Valley might appear rather chaotic, given the multiple flows of pedestrian and car circulation, the dramatic interaction of horizontal and vertical planes and the variety of architectural styles. And yet it bears the marks of its history: the French style of the Municipal theatre evokes the 1920s when São Paulo was the centre of the coffee economy; the first skyscraper, showing a mixture of modern technology and Italianate fashion, celebrates the success of Italian migrants; the American look of other buildings recalls the optimism of the 1950s, when Brazil switched into the idea of modernization and economic development; and a postmodern look keeps the city apace with present times.

(Andreoli, 1995, p.64)

FIGURE 2.5   *Canyon streets, São Paulo*

In one sense, then, the material mix of the city is still very much active through its symbolic contrasts, especially when you add to that the comings and goings of those who make up the Valley at different times of the day noted earlier. And indeed, many cities are like that:

- they are often places of extraordinary multiplicity, not to mention surreal or awkward in style

- where the juxtaposition of their different buildings is often a monument to times when a different set of rhythms, usually governed through events elsewhere, attempted to superimpose itself on a city.

Moreover, even when a particular phase of a city's formation has passed, the symbolic power of such monuments may still express an authority which no longer rests on any social or economic basis. It depends. It depends, as with rhythms and codings, upon the ability of social groups to impose and maintain their symbolism through the interpretation and use of the built environment.

ACTIVITY 2.3   This can be a very varied process, but if you turn to Reading 3A by the cultural geographer, Jane M. Jacobs, at the end of Chapter 3, you should be able to gain an insight as to what this can involve. Try not to dwell on the detail of the reading now: just skim-read the account and try to follow the recent struggles over how the monumentality of the City of London's landscape should be *re-presented* in these changing global economic times. The only issue that you should bear in mind is how the symbolic meaning of something as fixed as the built environment is itself far from settled. ◆

As I understood the piece, the core of the issue is the attempt to preserve an idea, a past rooted in the grandeur of the buildings and Empire, as an active memory which still has relevance and meaning today: how, in other words, it is possible to give such grand buildings as the Royal Exchange, the Bank of England and the Mansion House a contemporary symbolic significance and authority not only in an altered part of the City of London, but in a world that has altered London's financial role within it. Even though London is no longer the centre of an Empire, how should such a landscape be re-imagined to preserve a sense of authority in an entirely different global context without straightforward recourse to nostalgia? That such a task is attempted at all is, in many ways, a testament to the fact that buildings, in Thomas Markus's (1993) words, are 'a developing story', 'an unfolding event', 'the traces of which are always present'.

Indeed, as part of any such story, buildings can 'lose' their symbolic significance through transformation and use – as, for example, did many of the western buildings erected on the Bund waterfront in Shanghai when the full impact of the communist revolution was felt – only to see it reappear later in a different guise under a different set of relationships and connections. The Hong Kong

and Shanghai Bank, for instance, built in the heyday of western influence in
Shanghai in the 1920s (see Figure 2.6) – complete with granite exterior and a
series of elaborate internal mosaics depicting the Bank's worldly connections to
the likes of New York, Tokyo, Paris and London – was stripped of its global
meaning shortly after the 1949 revolution when the communist municipal
government moved in. Opting for cosmetic rather than structural change, the
communist party swapped the ornate crown on the domed roof for a red star
and plastered over the Bank's pictorial market connections and, in so doing,
deliberately erased the symbols of western monumentality. Come the 1990s,
however, on the back of the Chinese government's policy of 'opening up to the
west', the old Bank building has been restored to its former splendour
(including the preservation of the British made Royal Doulton toilets) and is
now occupied by Shanghai's flagship financial body, the Pudong Development
Bank. On the basis of such an institution, and indeed the many banks in the
Pudong economic zone, China hopes to exert an influence on contemporary
global financial affairs. In this new global context, the symbolism of past
connections has thus been reworked in an attempt to raise the profile of

Shanghai as a major
financial centre in East
Asia. Meanwhile,
further down the
Bund, the classic
Shanghai Club
building of the 1920s –
where once you had to
be white, male and
British to stand any
chance of membership
– now plays host to a
branch of Kentucky
Fried Chicken,
complete with its
plastic logo adorning
the neo-classical
entrance to the
building.

FIGURE 2.6
Mosaic mural in the
Hong Kong and
Shanghai Bank in
Shanghai

Bearing these examples in mind, the 'wildness' of cities, the evident changes and contrasts in the material and social make-up of cities, may not be so much of an exception, therefore, as something which, in Henri Lefebvre's terms, is routinely suppressed by those who attempt to speak for the whole. It may be easier to convey this point if we first consider the role that monuments and monumental spaces are said to perform in this process.

In *The Conscience of the Eye* (1970), one of a number of books by Richard Sennett on the design and social life of cities, he draws attention to the nature of the sombre monuments that dominate many a western city. Drawing a distinction between power and authority, he argues that particular kinds of monumental buildings – courts, museums, churches, various public and private institutions – represent 'spaces of authority'. They are less a mark of domination and power, and more a symbol of authoritative values and judgements. They exude a form of symbolic guidance and sense of order which is difficult to confront, in part because they often operate through closed doors or restricted access. We have seen this for example in the case of the Bank of England or the Mansion House in the City of London (see Reading 3A), or in relation to the classic buildings on the Bund in Shanghai in the 1920s. Authority in this context is thus primarily established visually: the splendour of the buildings, their intent to impress and, above all, their inaccessibility. This may well be compounded through the effects of light, colour or depth, although only in so far as such symbols reinforce their sense of detachment from the rhythms of daily life.

While not all monumental buildings or spaces operate through this kind of authoritative symbolism, all generate what Lefebvre refers to as a sense of 'membership'. This sense of who belongs and who does not is obviously intended to be exclusive. But this has little to do with closed doors and rather more to do with who is *recognized* as present. According to Lefebvre, this is achieved through an ability at any one point in time informally to prescribe what may and what may not take place in a particular building or in a particular location. So, for example, the buildings and spaces of a global finance house in New York or London or Frankfurt, or even the financial districts of which they are a part, if they are to be described as monumental, would effectively erase the traces of others in those spaces through a series of rituals, gestures, mannerisms, and other assorted practices. The monumentality of such spaces, in Lefebvre's sense of the term, would draw its meaning from the sounds, images, and even the scent of those recognized as 'members'.

Excluded, then, would be those whose rhythms and movements do not accord with the dominant interpretation and use of such spaces: in this case, much of the street life and its dealings, the routine service-workers who pass through the buildings unnoticed, the ancillary services (above and below ground) who keep the physical operations running, and so forth. Broaden the illustration to encompass many of the incongruities that Lévi-Strauss spoke about – palatial buildings next to impoverished dwellings, prestigious offices next to craft workshops (or even sweatshops) – and it is possible to see how much of city life can be filtered out through the lens of homogeneity. On this view:

● symbolic exclusion works against the notion that cities are places of 'wild' juxtapositions and serves to repress the plurality of cities in favour of an all-pervasive sameness.

Having said that,

● those excluded in one context may themselves work to define their own sense of membership and inclusive identity in another. Ordinary spaces can carry their own monumentality too.

This may sound rather curious, but the ever-present proximities to be found in cities can provide us with an example.

ACTIVITY 2.4   Aubervilliers is located in the northern half of Paris, at the end of one of the most working-class areas of the city. It is close to La Courneuve, Cité des 4000, an amalgamation of thin residential 'walls' built in the 1960s to accommodate those in some of the worst housing in the city (see Figure 2.7). The 'walls' or residential blocks have no authoritative symbolism like many a public building, nor – for some – any architectural merit. But they are monumental. And they are monumental to many of those excluded from membership in the more exclusive locations that the city has to offer.

Now read Extract 2.2 which is François Maspero's (1994) account of what the Cité des 4000 symbolized to those who lived there. ◆

EXTRACT 2.2
François Maspero: 'Cité des 4000'

The 4000 does not hold the absolute record for the longest 'walls' in France; apparently that goes to one 700 metres long built by B. Zerhfuss in Nancy: quite a feat. But still, the 4000 – four thousand flats for as many families, which makes how many inhabitants: 20,000? – is a fine example of human storage. It is one of the most grandiose results of the Delouvrier Plan. The year was 1960. 'Delouvrier', said de Gaulle, 'these inhuman suburbs are making the Paris area a shambles – sort them out'…

And then later, twenty years later, with Mitterand as president, they realized that the Plan wasn't working, that it was unbearable, and it was decided to *sort the place out* again. As the new urban planners had finally realized that everything stemmed from a lack of humanism, they looked for a human dimension. And as there was a risk of social discontent boiling over, they decided to dynamite the biggest 'wall' in Cité des 4000 Sud …

FIGURE 2.7   *Cité des 4000, Paris*

The big 'wall' needed only ten seconds to fall elegantly down. An on-site reception followed. 'Ten seconds to wipe out high-rise depression', 'The mistakes of the past', headlined the following morning's papers. On the site today there remains a vaguely grassy area and a melancholy little tree planted by the youngsters who were born there: they say that this little tree and this big empty space are all the roots they have left. Because those youngsters are still there – living in other 'walls', which get the sun now. Well, they're not all there, of course. Some of them had to go. The authorities also took the opportunity to 'ventilate' the immigrants. As they said, they're humanizing the place.

That 'wall', the absent, disappeared 'wall', was called Debussy...

In the end, maybe that is the hallmark of the 4000: its feeling of emptiness, even if it has the population of a small town. The feeling that no words exist to describe a giant 'development' which brings together and unites nothing, where nothing seems to have a meaning, not even that of a machine for living where nothing is attractive and nothing is ugly: where everything is a big dull zero. One dull 'wall' cancels out the next dull 'wall', and so on, from car park to car park, from paving stones to withered lawns; and nothing – but nothing – has an impact, so that out of so many zeros rises nothing but another zero just like all the rest. In the end, the only remarkable thing about the 4000 is the site of the demolished 'wall', the *cancelled* 'wall'. The 4000's youngsters are right to say, 'It's our monument.'

Source: Maspero, 1994, pp.156–61

What did you draw from Maspero's observations? Clearly there is a very different sense of membership involved here, where what is held on to symbolically is an altogether different world from the other monumental spaces described. There is little that is grand or authoritative about such spaces, yet they do serve as an identity for the youngsters on the estate. Even the grassy area created in the wake of the demolished wall, with its little tree, holds meaning for them. It is a part of their lives; a part touched by this surreal quality. But perhaps that is the point, and indeed part of the reason for the obvious sense of despair and incomprehension in Maspero's writing. The presence of the 4000 close to Aubervilliers, a short bus ride from the centre of Paris, is no more surreal in its contrast than many of those described by Lévi-Strauss some sixty years ago in São Paulo. It is in this sense that cities *are* 'wild' material places which no fixed image or string of representations can adequately capture.

3.2   CHARGED ENVIRONMENTS

There is perhaps more to this last sentence than we have thus far considered. Images, visual impressions, we have said, are insufficient in themselves to convey fully what it means to live in a city. We have to draw upon the full range of the senses, we have to 'take in' the city – its buzzing streets, its incongruous buildings, its 'giant' estates, the lot – through our bodies not just our eyes. But in Maspero's despairing tone and his projection onto the residential high-rise of a feeling of emptiness, where no words can possibly capture the meaning of the place, he is illuminating a different form of awareness that goes beyond that of sense impressions. He is drawing upon a form of *expressive* meaning.

What do we mean by this? Well, rather than being concerned with what we can see or hear in the city, expressive meanings have more to do with how the city is *felt* rather than how it is perceived or conceived. When Maspero writes about the 4000 – its residential 'walls' and its open spaces – as a development which 'brings together and unites nothing', 'a big dull zero', he is giving vent to his feelings, not his intellect. He was clearly 'moved' emotionally by the experience of visiting the estate and meeting those who lived there. The sense of futility in the air, the gloom attached to the development, are expressive in one way or another of what it means to live in such a place. The awareness is Maspero's. The awareness of someone from outside the estate, in this case, that of a Parisian writer and publisher. He was clearly 'struck' by what he had seen and heard, but his eyes and ears are only the media through which the place expressed its intensity for him. As we noted in the introduction, the city is as much 'out there' shaping our awareness of it, as it is endowed with meaning through our experience of it.

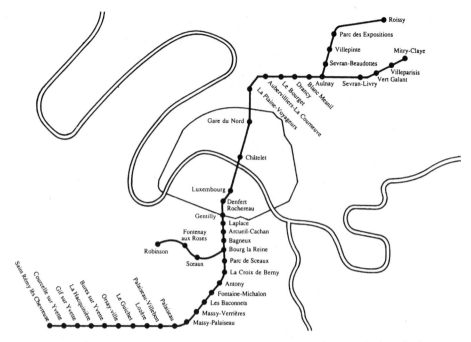

FIGURE 2.8 *Maspero's planned route on the Paris rail network (RER), from* Roissy Express

To place this example in context, Maspero, when he arrived at Aubervilliers, was on a journey through the suburbs of Paris which he intended to document with the help of a photographer, Anäik Frantz. Between them, they put together an account of their journey, *Roissy Express*, which built up an expressive awareness of life in the Paris suburbs. Rather than attempt to catalogue their experiences, they tried to trace the character of the different places as they found them:

- the different moods of the environments
- the intensities with which they were felt, and
- the almost surreal contrasts in the physical landscape.

In doing so, they called into question the limits of representation. They called into question how far concepts, words and images enable us to represent the worlds of a city in a meaningful way. Even photography is called to account:

> Exterior poverty, poverty for outside exhibition from the golden age of the picturesque – thank you Robert Doisneau, thank you Marcel Carné – is now only the fate of dropouts, tramps, drifters begging in the metro whom everyone more or less puts up with. But how do you photograph all the poverty behind the smooth walls, the silent walls – the poverty of depression and fear, of all the strains of everyday life, of so much loneliness?

(Maspero, 1994, p.190)

FIGURE 2.9
Exterior poverty, Paris

Not all of the journey through the suburbs of Paris is one of despair, however. Shades of joy run alongside gloom, agitation as well as pacification are equally present in their diverse encounters, and above all the built environment is shown to exert its own affects on those living in the various developments. Just as people can be awe-struck when confronted by the visual splendour of a particular building or setting, so they can be moved to despair by the relentless material conditions in which they live. This is not to suggest that everyone will experience the same material circumstances in exactly the same manner. On the contrary, the intensity with which people experience a particular environment is likely to be unique. However, what makes it possible to translate such experiences is the fact that such forms of expression are frequently shared or exchanged within a particular cultural context. They carry *meaning*.

Another way to come at this is to leave Paris for the moment and to consider once again Lévi-Strauss's enchantment with 1930s' São Paulo. Surreal landscapes, odd juxtapositions, incongruous settings are modes of expression used by Lévi-Strauss to communicate what struck him about the character of São Paulo on his first visit in the 1930s. To him, the garish colours of many of the buildings, the narrowness of the streets, the irregularity of building developments, and the manner in which the city laid itself out in an economic and social puzzle, are expressive of São Paulo at that time. While such descriptions cannot convey fully the atmosphere of daily life, the detail can point to wider juxtapositions of feeling. The palatial buildings mentioned earlier with their wrought-iron enclosures sealing them off from the rest of the city; the hillside shanty developments where the muddy torrents doubled as a source of water and as a sewer – the removed, cleansed nature of one, and the relentless nature of grime and dirt of the other. Each of these images carries meaning, but they are suggestive of more than symbolic representation; they are suggestive of a particular kind of awareness that is lived as much as conceived.

ACTIVITY 2.5   If you cast your mind back to your responses to Activity 2.1, these were perhaps closer to the kind of awareness that is more expressive than representational. You may recall that we spoke earlier about the moods of a city and, following Toni Morrison, drew attention to the sensual possibilities of city life, where in her words a city is capable of opening itself up to you. Think about that again as you read the following extract from *Jazz*. ◆

EXTRACT 2.3
Toni Morrison: 'Harlem, Springtime'

And when spring comes to the City people notice one another in the road; notice the strangers with whom they share aisles and tables and the space where intimate garments are laundered. Going in and out, in and out the same door, they handle the handle; on trolleys and park benches they settle thighs on a seat in which hundreds have done it too. Copper coins dropped in the palm have been swallowed by children and tested by gypsies, but it's still money and people smile at that. It's the time of year when the City urges contradiction most, encouraging you to buy street food when you have no appetite at all; giving you a taste for a single room occupied by you alone as well as a craving to share it with someone you passed in the street. Really there is no contradiction – rather it's a condition: the range of what an artful City can do. What can beat bricks warming up to the sun? The return of awnings. The removal of blankets from horses' backs. Tar softens under the heel and the darkness under bridges changes from gloom to cooling shade. After a light rain, when the leaves have come, tree limbs are like wet fingers playing in woolly green hair. Motor cars become black jet boxes gliding behind hoodlights weakened by mist. On sidewalks turned to satin figures move shoulder first, the crowns of their heads angled shields against the light buckshot that the raindrops are. The faces of children glimpsed at windows appear to be crying, but it is the glass pane dripping that makes it seem so.

Source: Morrison, 1993, pp.117–18

Morrison's style is personal and evocative. In the extract, she is trying to convey something of the sensuality of Harlem's social and material environment at a particular time of the year. The awareness she is building is nonetheless not that far removed from what both Maspero and Lévi-Strauss had in mind when they attempted to communicate cities as 'wild', diverse places. For each of these authors, in various respects, a fuller understanding of cities can only really be grasped through

● the charged, expressive meanings by which cities are presented to us as animated environments

● the felt intensities that enlarge the horizons of our sense-experience.

4 *City life: proximity and difference*

Up to now we have been concerned to show that cities are evocative places, places where people are drawn into all kinds of proximate relationships, often by chance, often fleetingly and often on an unequal basis. The social and built environment of the city as a series of odd juxtapositions, contrasting rhythms and symbolic meanings provides a setting for how different groups of people are drawn into proximity and how their worlds may touch. In many instances, the differences remain unacknowledged or even repressed to give a singular account of cities. Where the worlds do meet, however, especially in relation to where people live, the negotiation of difference and plurality can have quite different outcomes – and arouse quite different intensities.

- One outcome, as indicated in the previous chapter, is that people reject difference and draw 'walls' around themselves and their communities, either to defend themselves or, more commonly, to protect their advantages. This is the point at which physical boundary lines are drawn across a city in an attempt to fix its spaces.

- Another, more fluid possibility, however, is for people to remain indifferent to the close proximity of others, remote in their ways despite their nearness to everyone.

These are the two broad headings under which we will consider how the negotiation of difference has been approached and debated. Needless to say, they fail to capture all possibilities, but they do offer a clear way into thinking about city life and difference.

4.1 CLOSED WORLDS AND HIGH WALLS

In a book written about personal identity and city life, *The Uses of Disorder* (1971), Richard Sennett first coined the notion of 'purified communities'. The notion is an interesting one because in many ways it sums up how particular groups build 'walls' around themselves, sometimes literally, to exclude those who are not 'the same'. If, up to now, we have largely considered the cross-cutting nature of city life, then the notion of a purified community represents a block on diversity and a clear limit to the movement of 'others'. In short, difference is firmly rejected in favour of sameness. At one level, the idea of spatial separation explicit in this form of purity can be understood as a reaction to the extraordinary mix of city life, that is, as an inability to deal with the 'strangeness' of others. When the edges of different worlds meet, the experience is, quite simply, too much to bear and a form of anonymity is sought. But the formation of closed communities may also come about through more than acts of self-preservation. The construction of a group identity around who 'we' are, something that is held in common, can also be the hallmark of intolerance.

The implications of this view, where people close in on themselves by excluding 'outsiders', are potentially far-reaching for an understanding of how city space is divided. It takes us closer perhaps to what many would understand by the idea of worlds within cities: the spatial segregation of city life. When segregation takes on a spatial form, the marking of difference takes the form of boundary lines etched in city space. Here a real sense of juxtaposition may take the line of a street or an expressway, with one side divided off from the other by cultural and economic barriers. Street signs and the architectural style of buildings may both serve as symbolic boundaries which are often enforced by lines of access based on cost, which separate the wealthy from the less-than-wealthy, and so on right down to the poorest of estates. Many of the incongruities of urban life that we have spoken about, the plush residences next to impoverished housing, the different economic and social groups who dwell alongside one another, and so on, reflect the unequal distribution of resources that enables the more powerful groups to insulate themselves from others close by.

Not all differences, however, can be mapped according to the unequal distribution of resources. Cultural separation, the act of self-enclosure, may come about through groups seeking to protect themselves from the more economically or politically powerful. They seek the protection of their 'own' to avoid further discrimination or economic loss. Seeking solidarity within their own mutual support networks is a characteristic of rich and poor communities alike, but in the case of the latter the process of 'walling in' is likely to be a matter of necessity, not of choice.

The issue of choice is a critical factor in all this, for without it the notion of 'purified' communities is somewhat less than convincing. The ability to *impose* boundaries, to *limit* who may come and go within specific areas, is central here. In its starkest form, such forms of closure are illustrated by what have become known as 'gated communities'. A US phenomenon which sprang up in the 1970s, gated communities are controlled, private spaces which people have bought into for their own exclusive use. They are, literally, walled communities: housing built behind walls, designed to focus inwards so that adjacent street life is erased from the urban imagination. To ensure that only the 'right kind of people' are admitted, cost is obviously a major filter, but on an everyday basis control over movement is exercised through surveillance cameras, uniformed guards and the like. The 'outside' world of difference is effectively excluded and the movement of those who enter to serve those 'inside' is scrupulously monitored. It is all highly ordered and as a form of spatial segregation such communities offer those who live within them a special sense of membership.

FIGURE 2.10 *A secure residential enclave in São Paulo*

Gated communities are not the exclusive preserve of US cities (where some 8 million people are now said to live behind such walls). Variations on this enclave pattern are to be found in many cities which exhibit polarized spatial segregation. They have long been a feature of colonial cities, albeit with different surveillance and control techniques, and cities such as Istanbul, Mexico City, Jakarta and São Paulo boast their equivalent today.

ACTIVITY 2.6   Now read the following extract by Teresa Caldeira on the 'fortified enclaves' of São Paulo and take note of any rules of inclusion and exclusion that endow membership to these residential communities

Before you do so, however, you will find it useful to remind yourself of Henri Lefebvre's sense of membership outlined in section 3.1. In particular, you should consider how a sense of closure is achieved and monitored, and at what cost to those involved. ◆

EXTRACT 2.4
Teresa Caldeira: 'Fortified enclaves: the new urban segregation'

São Paulo is today a city of walls. Physical barriers have been constructed everywhere – around houses, apartment buildings, parks, squares, office complexes, and schools. Apartment buildings and houses which used to be connected to the street by gardens are now everywhere separated by high fences and walls, and guarded by electronic devices and armed security men. The new additions frequently look odd because they were improvised in spaces conceived without them, spaces designed to be open. However, these barriers are now fully integrated into new projects for individual houses, apartment buildings, shopping areas, and work spaces. A new aesthetics of security shapes all types of constructions and imposes its new logic of surveillance and distance as a means for displaying status, and is changing the character of public life and public interactions …

Closed condominiums are supposed to be separate worlds. Their advertisements propose a 'total way of life' which would represent an alternative to the quality of life offered by the city and its deteriorated public space. The ads suggest the possibility of constructing a world clearly distinguishable from the surrounding city: a life of total calm and security. Condominiums are distant, but they are supposed to be as independent and complete as possible to compensate for it; thus the emphasis on the common facilities they are supposed to have which transform them into sophisticated clubs. In these ads, the facilities promised inside of closed condominiums seem to be unlimited – from drugstores to tanning rooms, from bars and saunas to ballet rooms, from swimming pools to libraries …

The middle and upper classes are creating their dream of independence and freedom – both from the city and its mixture of classes, and from everyday domestic tasks – on the basis of services from working-class people. They give guns to badly paid working-class guards to control their own movement in and out of their condominiums. They ask their badly paid 'office-boys' to solve all their bureaucratic problems, from paying their bills and standing in all types of lines to transporting incredible sums of money. They also ask their badly paid maids – who often live in the *favelas* on the other side of the condominium's wall – to wash and iron their clothes, make their beds, buy and prepare their food, and frequently care for their children all day long. In a context of increased fear of crime in which the poor are often associated with criminality, the upper classes fear contact and contamination, but they continue to depend on their servants. They can only be anxious about creating the most effective way of controlling these servants, with whom they have such ambiguous relationships of dependency and avoidance, intimacy and distrust.

Another feature of closed condominiums is isolation and distance from the city, a fact which is presented as offering the possibility of a better lifestyle. The latter is expressed, for example, by the location of the development in 'nature' (green areas, parks, lakes), and in the use of phrases inspired by ecological discourses. However, it is clear in the advertisements that isolation means separation from those considered to be socially inferior, and that the key factor to assure this is security. This means fences and walls surrounding the condominium, guards on duty twenty-four hours a day controlling the entrances, and an array of facilities and services to ensure security – guardhouses with bathrooms and telephones, double doors in the garage, and armed guards patrolling the internal streets. 'Total security' is crucial to 'the new concept of residence.' Security and control are the conditions for keeping the others out, for assuring not only isolation but also 'happiness,' 'harmony,' and even 'freedom.' In sum, to relate security exclusively to crime is to fail to recognize all the meanings it is acquiring in various types of environments. The new systems of security not only provide protection from crime, but also create segregated spaces in which the practice of exclusion is carefully and rigorously exercised …

The characteristics of the Paulista enclaves which make their segregationist intentions viable may be summarized in four points. First, they use two instruments in order to create explicit separation: on the one hand, physical dividers such as fences and walls; on the other, large empty spaces creating distance and discouraging pedestrian circulation. Second, as if walls and distances were not enough, separation is guaranteed by private security systems: control and surveillance are conditions for internal social homogeneity and isolation. Third, the enclaves are private universes turned inwards with designs and organization making no gestures towards the street. Fourth, the enclaves aim at being independent worlds which proscribe an exterior life, evaluated in negative terms. The enclaves are not subordinate either to public streets or to surrounding buildings and institutions. In other words, the relationship they establish with the rest of the city and its public life is one of avoidance: they turn their backs on them. Therefore, public streets become spaces for élite's circulation by car and for poor people's circulation by foot or public transportation. To walk on the public street is becoming a sign of class in many cities, an activity that the elite is abandoning. No longer using streets as spaces of sociability, the élite now want to prevent street life from entering their enclaves.

Source: Caldeira, 1996, pp.307–14

One of the things that struck me about this extract was the extent to which conformity and order is formally secured in these communities. It is secured not simply, as in Lefebvre's view, through the recognition or approval of a dominant way of doing things, but through formal rules of conduct. The enclaves described by Caldeira are very much *single-minded spaces*: spaces which require those who live there to conform to an imposed sameness. There is none of the ebb and flow of Jacobs' street life, its chance encounters, or its amiable strangeness. And there is certainly none of the 'wildness' or incomprehension of city life that Maspero discovered on his urban travels. What we are confronted with in this type of enclave are rules which forbid difference, almost of any kind – in many cases down to what colour the dwelling and its doors should be, what shrubbery in the garden is allowed, and what time of the day noise is appropriate, and so on. Membership of such communities is drawn so sharply that those on the 'inside' are left in little doubt as to what their enclosed space represents. Such rules dramatize the social distance between those who live in these communities and those who are on the outside. They say something that is less than positive about the worlds outside. More importantly, the boundary line between the inside and outside of such communities does more than close one group off against others. It also *connects* them, by virtue of the differences drawn. Let me elaborate.

When people draw 'walls' around themselves in this way, their sense of who they are is not established in a vacuum. It is established *in relation to* how they think others live and behave. The idea of living behind a three-metre high wall with your 'own' kind in São Paulo, or wherever, tells us as much about those living on the other side of the wall as it does those who choose to retreat behind their residential 'walls'. If the enclosed community imagines itself to be respectable, civilized, law-abiding, middle-class, and the like, then this is precisely because it imagines that many of those on the outside are *not* these things. In this way, the insulated community establishes a sense of itself from that which it is not.

In such circumstances, a different kind of intensity may be felt by those 'walled in': one of insecurity or fear which may manifest itself through an intolerance of others. This is not to suggest that all differences in a city are felt in quite this way or that they always take the form of a hard, spatial boundary. Rather it is to say

- that when differences are negotiated negatively in the city in this manner, one outcome is a form of segregation and exclusion which can reinforce existing social and economic inequalities

- that the power of certain groups defines and limits who mixes, in much the same way that earlier we spoke about powerful groups in the city superimposing their rhythms on others.

In the context of residential enclaves, however, that difference and diversity is not so much smothered as physically barred.

4.2 INDIFFERENT WORLDS AND DETACHED LIFESTYLES

Residential segregation is one of the sharpest ways in which the different worlds of a city are produced and 'gated communities' are possibly their most extreme expression. However, much of city life, as we have seen, is not divided along such stark lines and difference is not always negotiated in such dramatized ways. City worlds are invariably more jumbled than that, with all kinds of overlay, displacement and crossings entering into their composition. In *The Uses of Disorder,* Sennett goes on to extol the virtues of the tangled movements which characterized Chicago's migrant communities in the early part of the twentieth century, after setting out his abhorrence of purified communities. Fast disappearing, according to Sennett, the threaded movements and rhythms of daily life described in Chapter 1, which saw Polish, Irish, Greek and Italian migrants cross the paths of one another, worked against the possibility of community closure. As he saw it:

> They *had* to make this diversity in their lives, for no one of the institutions in which they lived was capable of self-support. The family depended on political 'favours', the escape valve of the coffee shops and bars, the inculcation of discipline of the *shuls* and churches, and so forth, for ongoing support. The political machines tended in turn to grow along family lines, to interact with the shifting politics of church and synagogue. This multiplicity of contact points often took the individuals of the city outside the ethnic subcultures that supposedly were snugly encasing them ... This multiplicity of contact points meant that loyalties became crossed in complex forms.
>
> (Sennett, 1970, p.56)

For Sennett, this diversity is now no longer made to work. The contact points have been eroded by institutional change and where there is diversity in the city, it 'does not prompt people to interact' (Sennett, 1994, p.357). The retreat to closed communities is a feature of many cities and, where difference is apparent, the proximity of different groups has not produced Jane Jacobs' ideal meeting-place, but rather a setting of indifference.

This is clearly a depressing scenario for Sennett and, as we have seen in section 2, the daily rhythms which compose much of city life can lead to contexts where indifference and a lack of acknowledgement are lived uneasily by certain groups. But the recognition of social difference within cities does not necessarily lead to mutual withdrawal, as Sennett presupposes. Differences and diversity, as Iris Marion Young has argued in *Justice and the Politics of Difference* (1990), may well involve the possibility of incomprehension and misunderstanding between groups in the city, yet that is not to suggest that meaningful interaction across differences of culture and ethnicity is unlikely. Because cities are places where all manner of different people live in close proximity, their relative 'strangeness' to one another may simply be a condition of city life. In other words rather than close themselves off to one another, different kinds of people may negotiate their social distance from others as part of going about their daily business and coping with whatever comes their way.

Even though they may share the same spaces and facilities in the city, therefore, they may do so without there being a felt need to assert their difference. Indeed, the possibility of co-existence may actually depend, as Simmel (1903/1950) has argued, on developing *mutual forms of distancing.*

Indifference in this context, then, is not so much a relation of power at close quarters, as one of mutual apprehension. The previous chapter touched upon this in relation to Wirth's account of the intensification of social interaction in the city and the need for people to immunize themselves against the expectations of others. But if left to take its own course, a lack of understanding or a miscommunication may well develop into mutual recognition, rather than mutual withdrawal. Here is Maspero again – in Extract 2.5 – a little further on in his journey across Paris, at the Plaine Saint-Denis, close to one of the capital's main railway stations, the Gare du Nord. What is interesting about the extract is the way that both apprehension *and* recognition feature in the encounters.

EXTRACT 2.5
François Maspero: 'The Plaine Saint-Denis'

The alley comes out on to straight streets running up against the embankment where the RER goes past; on either side, tiny brick *pavillons.* One of them is a tarnished marble plaque on which François makes out:

> *Here lived* **MARIA RUBIANO**
> *who died at Ravensbrück*
> *1944*

He asks Anaïk to photograph it. A couple comes out and asks what they are doing. 'Are you from the town hall?' A strong Portuguese accent. They are visibly worried. For them, it seems, any stranger who shows a remotely close interest in their street can only come from the town hall, and everything from the town hall means inventories of fixtures, property deals and rehousing, therefore departure, if not eviction. The travellers explain. It's not easy: diffuse embarrassment and fear hang in the air. François talks about his interest in the plaque. Did they know Maria Rubiano? No, they arrived later. A sad story, says the man:

'It was a woman who lived in the house. There was an air raid – she got out but died instantly.'

Anaïk says how she likes this quiet street at the world's end. She is sincere, and they believe her. Photo-time in front of the plaque, their grandson in

FIGURE 2.11   *The Plaine Saint-Denis, Paris*

their arms. … Further along, opposite a Portuguese café, Anaïk
photographs Madame Pauline's caravan. Again the fear is tangible:
Madame Pauline comes out and has to be reassured that they're not from
the town hall, that they're not after her dogs, that they're not there to get
her out. They end up having a beer together in a café whose customers,
gathered at the bar, listen and watch in silence. Anaïk will come back with
the photo. It is the start of a friendship.

The only shops here are abandoned, except a chemist's. There's a Spanish
church in reinforced cement, its locked-up hostel oozing rust. The hostel
opens at weekends. They will be coming again: some people there speak
Castillian, though Galician and Portuguese are more common and Cape
Verde creole even more so, besides various African dialects. There is a
warm atmosphere, you can eat tapas and greasy fried cod, drink San
Miguel beer and play extremely lively games of dominoes.

In the late afternoon, Plaine Saint-Denis emerges from its torpor. The
children play freely in the quiet streets. Where in Paris can the children
still play in the street? The children of the Plaine Saint-Denis are beautiful,
like those from Blanc-Mesnil and Les Beaudottes. The Plaine offers an
image of a world coming apart, but its inhabitants, who live so badly, are
hanging on to life with a vengeance. Many recent arrivals come from the
Cape Verde Islands: the Cape Verdian people have some of the most
harmonious features in the world: on those volcano splinters scattered a
thousand kilometres off the coast of Africa, over several hundred years, all
the African races have blended together with those the Portuguese
snatched from beyond the Indies …

> Rue du Landy runs directly north between deserted warehouses and overcrowded houses as far as the Canal Saint-Denis. You leave 'Little Spain' and 'Little Portugal' behind. The closer you get to the canal, the more it seems most people are becoming North African. Narrow housing shared by several families gives way to furnished hotels above cafés which still have names from yesteryear: 'L'Embuscade' ('The Ambush'); or others which have received new ones, such a 'L'Oasis'. Here suspicion is heavy in the air. Slum landlords. Recent illegal immigrants. Drug trafficking. Let us move on.
>
> Source: Maspero, 1994, pp.207–9

Here, then, we have the different worlds of a city spilling over into one another spatially and socially. The 'outsider' in this context is 'the stranger' from the town hall, who is feared simply because the residents are frightened that they will lose even more than they have already. Once the barriers of misunderstanding have been breached however, openness rather than closure characterizes these encounters, without a denial of difference. Of course, social interaction in the city is not always like this, but the example does warn against any broad claims that social diversity prohibits such forms of interaction.

The example also alerts us to an awareness of difference that does not involve rejection as a response. Perhaps Simmel was indeed correct when he observed that, in the city, people distance themselves from others in close proximity principally because they have to co-exist. The social reserve, the air of detachment, the general indifference, are in that sense simply means of coping with city life and difference, not a denial of it. The need to preserve a degree of anonymity was, for Simmel, a crucial feature of urban lifestyles. At the core of his thinking was the assumption that city life, with its overbearing rhythms and its many unscripted encounters, was simply more than most people could possibly be expected to bear. Some form of reserve and detachment of feeling was a necessary feature of urban interaction. While he may have exaggerated the point, the ambivalent experience that Simmel (1908b/1950) described in relation to social interaction – as a mix of remoteness and proximity – conveys something of the intensity that cities can arouse. If so, then a sensitivity to living city life at a distance may not be quite the depressing scenario that Sennett suggests, but rather a necessary condition of what it is *to be* in the city.

5 *Conclusion*

Throughout this chapter, our concern has been to explore the many worlds of the city as they are drawn together in close proximity. So far as cities are concerned, it is the close proximity of many others in all manner of arrangements and relationships which give the urban its distinctive feel and presence. Whether São Paulo or Shanghai, Nairobi or Paris, cities bring together different worlds in diverse and often surprising ways: through the constant and successive rhythms by which people move in and across one another, through the displays of indifference which pass for the negotiation of difference and, more pointedly, through the construction of high walls which serve both as a barrier between the different worlds and as a connecting link.

We have also been concerned to show that the terms by which these worlds negotiate their differences or by which they slide across one another involves wider juxtapositions of power and of meaning. The ability to smother difference or to erase the presence of another's space, as Lefebvre would have it, or to construct the kind of single-minded spaces behind enclosed walls that Caldeira outlines, is a reminder of the differential ability of groups in the city to divide up urban space and, often routinely, to impose a dominant coding upon it. It is also a reminder, however small, that power is rarely practised with complete success, as the ability of the Parisian youngsters to construct their own 'monumental' spaces on an outer suburban estate testifies. Ordinary spaces, as much as authoritative spaces, are subject to the same degree of contestation and feeling.

All this, in fact, points to the significance accorded in the last chapter to the *intensity* of city life. When large numbers of people live, work and get by in close proximity, often in quite ordinary spaces, the differences of culture, of rhythm, of power, are often *felt intensively*, rather than simply observed and passively recorded. This expressive side to city life, where people can be moved by the sheer physicality of the city and its symbolic contrasts, as well as its rhythmic movements and everyday encounters, is what this chapter adds to our understanding of cities as intense environments.

In the next chapter, we take this understanding a step further by considering the intensity of city life in the wider context of the relationships between cities.

References

Andreoli, E. (1995) 'The visible cities of São Paulo' in Borden, I., Kerr, J., Pivaro, A. and Rendell, J. (eds) *Strangely Familiar: Narratives of Architecture in the City*, London, Routledge.

Caldeira, T. (1996) 'Fortified enclaves: the new urban segregation', *Public Culture*, vol.8, pp.303–28.

Gilsenan, M. (1982) *Recognizing Islam*, New York, Pantheon.

Gugler, J. (ed.) (1997) *Cities in the Developing World: Issues, Theory and Policy*, Oxford, Oxford University Press.

Jacobs, J. (1961) *The Death and Life of Great American Cities*, New York, Random House.

Jacobs, J.M. (1996) *Edge of Empire: Postcolonialism and the City*, London, Routledge.

Lefebvre, H. (1991) *The Production of Space*, (trans. D.Nicholson-Smith), Oxford, Blackwell.

Lefebvre, H. (1992) *Élements de Rythmanalyse: Introduction à la Connaissance des Rythmes,* Paris, Syllepse-Périscope.

Lefebvre, H. (1996) *Writings on Cities*, (selected, trans. and intro. by E.Kofman and E.Lebas), Oxford, Blackwell.

Lévi-Straus, C. (1973) *Tristes Tropiques*, London, Jonathan Cape.

Lomnitz, L. (1997) 'The social and economic organization of a Mexican shanty town' in Gugler, J. (ed.).

Mai, U. (1997) 'Culture shock and identity crisis in East German cities' in Öncü, A. and Weyland, P. (eds).

Markus, T. (1993) *Buildings and Power*, London, Routledge.

Maspero, F. (1994) *Roissy Express: A Journey through the Paris Suburbs*, London, Verso.

Melbin, M. (1987) *Night as Frontier: Colonizing the World After Dark*, New York, The Free Press.

Morrison, T. (1993) *Jazz*, London, Picador.

Nelson, N. (1997) 'How women and men get by and still get by (only not so well): the gender division of labour in a Nairobi shanty-town' in Gugler, J. (ed.).

Öncü, A. and Weyland, P. (eds) *(1997) Space, Culture and Power: New Identities in Globalizing Cities*, London, Zed Books.

Sassen, S. (1989) 'New York City's informal economy' in Portes, A., Castells, M. and Benton, L.A. (eds) *The Informal Economy: Studies in Advanced and Less Developed Countries,* Baltimore, MD, The Johns Hopkins University Press.

Sennett, R. (1970) *The Conscience of the Eye: The Design and Social Life of Cities*, London, Faber and Faber.

Sennett, R. (1971) *The Uses of Disorder: Personal Identity and City Life*, London, Allen Lane.

Sennett, R. (1994) *Flesh and Stone: The Body and the City in Western Civilization*, London, Faber and Faber.

Simmel, G. (1903) 'The metropolis and mental life' in Wolff, K.H. (ed.) (1950).

Simmel, G. (1908a) 'Sociology of the senses: visual interaction' in Park, R.E. and Burgess, E.W. (eds) (1969) *Introduction to the Science of Sociology* (3rd edn), Chicago, IL, The University of Chicago Press.

Simmel, G. (1908b) 'The stranger' in Wolff, K.H. (ed.) (1950).

Toth, J. (1993) *The Mole People: Life in the Tunnels Beneath New York City*, Chicago, IL, Chicago Review Press.

Weyland, P. (1997) 'Gendered lives in global spaces' in Öncü, A. and Weyland, P. (eds).

Wolff, K.H. (ed.) (1950) *The Sociology of Georg Simmel*, New York, The Free Press.

Young, I.M. (1990) *Justice and the Politics of Difference*, Princeton, NJ, Princeton University Press.

CHAPTER 3
Cities in the world

by Doreen Massey

1 *Cities interlinked*

1.1 THE SQUARE OF THE THREE CULTURES

Not far from the centre of today's Mexico City there is a square called *La Plaza de las Tres Culturas* – the square of the three cultures. It is really just a part of the city in which are exposed to view in immediate proximity elements of the three major cultures which have gone in to making this place. Three cultures juxtaposed. There are the excavated ruins of an enormous Aztec pyramid. There is a seventeenth-century baroque Roman Catholic church. And there is a complex of buildings in the International Style. The pre-Columbian, the Hispanic-colonial, and the modern.

La Plaza is a moving and impressive monumental space. The pyramid, now just a vestige of its previous self, somehow silent and unreachable; the church (built, in a defiant gesture, right up against the razed pyramid) still used but now showing its age; and the (relative) newness and shininess of the more recent constructions. Together they constitute a monument which makes you think. How is it to be understood, what is it trying to say?

FIGURE 3.1 *La Plaza de las Tres Culturas*

On the one hand, it is possible to read the monument as a reminder of a lost past. The devastation of the pyramid and the physical assertiveness of the church in its positioning are certainly emblematic of the destruction this city underwent on the arrival of the Spanish. Physically and culturally much was lost. Perhaps, then, the monument can be read as a burrowing back through the layers of time in search of the city's roots.

Or one could interpret it another way. Perhaps the monument is rather a bringing together of all the elements which still today (and however unequally) make up this city. For while the buildings certainly succeeded each other, the social forces and cultures which they represent each still have a presence here. Not, then, so much a search for roots as a celebration of the mixture which is this place. It is this second interpretation which coincides more closely with the intentions of those who created this project in the 1960s.

ACTIVITY 3.1   Go back to section 3.1 of the previous chapter and re-consider what is being said there about buildings and especially about monumental buildings.

These buildings in *La Plaza de las Tres Culturas* bring together periods of the city's history just as does the mixture of buildings in São Paulo's Anhangabaú. They reveal elements of the city's 'different histories'. But this square is not just a jumble of city buildings from different periods; it is preserved and designed as a monumental space. What was it Lefebvre argued about the function of monuments? ◆

Lefebvre argued that monuments have a function of establishing membership. His view was that 'monumental space' offers 'each member of society an image of that membership', that it constitutes 'a collective mirror' (Lefebvre, 1991, p.220). As it says in Chapter 2, what is at issue is *recognition*.

When *La Plaza de las Tres Culturas* was constructed (that is, when the modern buildings were built and the unity of the three periods/cultures was bound into the design) it was precisely with this kind of message in mind. On the one hand it was a physical recognition of the contribution of indigenous and Hispanic cultures to the Mexico of today. And on the other hand it was an invitation to all Mexicans – indigenous, mestizo, white – to recognize themselves in, as part of, the modern city. This, then, is a monument which recognizes the city of today as a mixture evolved over time.

It is, moreover, a mixture in a geographical sense as well. The monument does not speak of this explicitly, yet it is there in the very structures which form it. Thus, the modern buildings share much with such buildings all around the world. You might see their like in Africa or in Asia. The church is reminiscent in its architectural style of Spanish Catholicism in Europe. Neither architecture is simply and authentically 'of this place'. Both, in their physical construction, express *both* the specificity of this place *and* links with the world beyond. This composite nature (of local and global, one might say), moreover, can be detected in the pyramid too. Without doubt a temple of the local Aztecs, the architecture reflects also the traditions of the wider region of mesoamerica. Each of the 'elements', the societies and cultures, which has gone into making Mexico City what it is today was itself tied into a wider geographical set of interconnections. Each brought together elements from near and far to create its own specificity.

The city is, then, not only a developing mixture through history, but also in each moment of that history, *the focus of a wider geography,* bringing together differences in space. Cities are foci of changing patterns of interconnections.

The aim of this chapter is to consider cities in the context of these wider geographies. It will draw back a little from the exploration of the internal intensities of individual cities (though we shall keep an eye on these too) in order to develop an approach to cities within their wider sets of interrelations.

Even this brief consideration of *La Plaza de las Tres Culturas* has enabled us to understand Mexico City as both

● a developing mixture through history, and also

● in each moment of that history the focus of a wider geography, bringing together differences in space.

If you want to put it at its most abstract, what I am proposing is the city as an intense focal point or a node of social relations in time and space. And this is how we shall interpret cities in this chapter.

1.2   A BIT OF HISTORY

When the Spanish arrived in this city (in their year of 1521), it really was a meeting of two worlds. Indeed, in these supposedly more enlightened times it has been re-named: from *la conquista* – the conquest – to *el encuentro* – the meeting. It was the coming-together of two cultural histories which up until that point had had no knowledge of each other. They had been separated by an ocean across which ran no current social or cultural connections. Geography had kept the two histories apart, their stories had remained separate. But now they were to meet.

Each, indeed, began with massive misconceptions about the other. The Spanish were part of that European exploration westwards led by a Columbus who firmly believed he was on the way to India. He and those first explorers had no notion that a whole other continent lay across their path. And for their part, the Aztecs completely misinterpreted the arrival of these fair-skinned men on strange animals. For them, this was not the mid-sixteenth century: when the Spanish had first landed on the Yucatán coast far to the east of the city (in Spanish year 1519) it was to the Aztecs the year One Reed in their 52-year cycle. And that was a year

FIGURE 3.2 *The island capital of Tenochtitlán (as imagined and painted by Miguel Covarrubias, 1904–1957)*

of significance: it was the year in which the deified king Quetzalcoatl had disappeared towards the east (was he therefore about to return, was the news coming in of these strange landings a warning of his imminent arrival?) and it was also a year of special significance for the cosmological sign of the plumed serpent – of Quetzalcoatl himself. This too was ominous. There has been much academic debate about the significance of these temporal coincidences for the Aztecs. But there seems to be agreement that, as Moctezuma puzzled over the news coming in from the east, it all added to his uncertainty and sense of foreboding.

As they rode down into the wide and spreading Valley of Mexico in which lay the Aztec city of Tenochtitlán, the Spanish were stunned (see Figure 3.2). They wrote afterwards about their complete amazement. For here was a city equal to anything in Europe: in size, in social complexity, and in built form. The central square was a huge complex of massive stone buildings: pyramids, ball-courts, temples, palaces. The market to the north of the centre, at Tlatelolco where *La Plaza de las Tres Culturas* now stands, attracted some 25,000 people every day, and between 40,000 and 60,000 on the days of special markets (once every fifth day) (Townsend, 1993, pp.173–4). There was a complexity of agriculture, craftwork, trade and military activity. This was an accomplished imperial capital.

Many of the local products were completely unknown to the Spanish, while the Aztecs had never seen horses before. The two cultures had different concepts of space and time. They valued different things. In material terms the Aztecs valued jade and turquoise and this is what they offered the Spanish when the latter demanded precious goods. As Vaillant (1944/1950, p.133) puts it, 'Such misguided compliance was highly irritating to Cortés and his men'! What the

Spanish were after –
what *they* valued – was
gold. But gold was only
valuable to the Aztecs
for the ornaments which
could be made of it; for
them it was not the
foundation of an
exchange and currency
system, and they
remained puzzled.

This, then, really was
(the beginning of) the
meeting of two worlds.
Two separate histories,
up to now operating in
completely separate
spaces, met up with each
other in the first half of
the sixteenth century.

But, even if we now call
it a 'meeting', this was a
coming-together of two
highly unequal powers.
The Aztec city was

virtually razed to the ground, and a new built environment was created. The
baroque church was built where once stood the pyramid at Tlatelolco, and the
present Spanish main square, with cathedral and national palace, was erected
where once had been pyramids and ball-courts and temples. Indeed, they were
not only built 'in place' of the Aztec city, but – it is thought – they probably used
the same materials: the masonry of the pyramid, or temple, was taken to build the
church. One city centre almost obliterated the other. Only the occasional Aztec
building remained among the devastation. The physical city was easier to destroy
than the social and the cultural. But the Spanish – or some of them, and at first –
did their best. Records were destroyed, cultural habits disrupted, even concepts
of space and time were obliterated by the imposition of the new Spanish ways.
And almost as completely as the material city, something else was destroyed: the
city as an Aztec centre of a wide and complex geography of cultural
interconnections, trade and military power.

If, as Chapter 1 argued, one of the things which is crucial to a city's creativity and
power is its position as a focus of social relations, then the Aztec city in this sense
was also destroyed at this time. Tenochtitlán was a focus both of trade and of

FIGURE 3.3
The story of the conquest of Mexico in indigenous symbols. On the left the symbol of One Reed (1519) and a Spaniard on horseback being met with a tribute of gold beads, and Cortés sits in the temple of Tenochtitlán; on the right, under the symbol of Two Knife (1520) Alvarado massacres the Indians at the great temple. (Source: Codex Vaticanus A)

empire. Its network of trade routes spread over all of what is now Mexico, and beyond, and the empire extended from the Pacific Ocean to the Gulf of Mexico. As part of the acknowledgement of imperial relations, surrounding areas sent in to Tenochtitlán tributes and levies of foodstuff, raw materials, and manufactured goods such as clothes and pottery. The empire and imperial relations were destroyed by the Spanish conquest. And within only five years of that conquest the long-distance trade network had also disappeared, a fact explained by Townsend (1993, p.186) as due to the network's dealing in importing luxury items – the feathers of tropical birds, greenstones, and exotic animal hides – which had high value for the Aztecs but not for the Spanish. From being the dominant focus of trade and imperial connections across mesoamerica, the city turned towards Madrid. Mexico City was installed as the capital of what now became New Spain (a part of the wider Spanish colonies in the Americas), but its local dominance was now in turn subordinated to an even greater power, a new imperial capital across the Atlantic. Compare maps (a) and (b) in Figure 3.4. In terms of its positioning within wider networks of social relations, the meeting of two worlds in this city – the passage from Tenochtitlán to Mexico City – meant a complete reorientation.

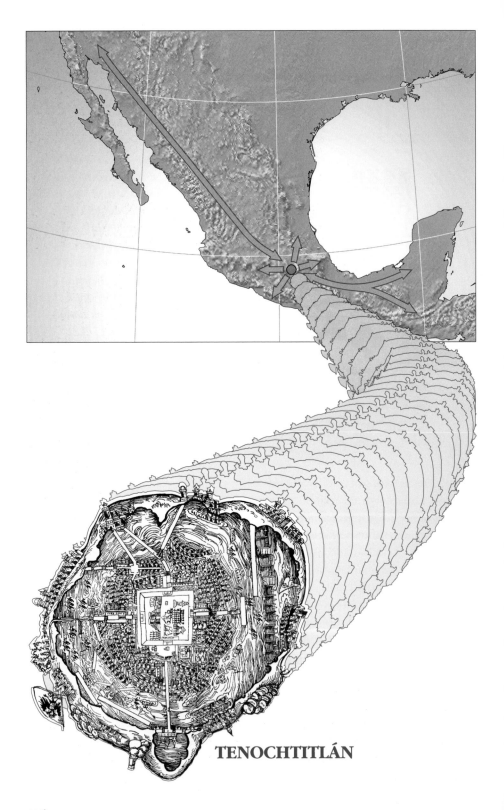

TENOCHTITLÁN

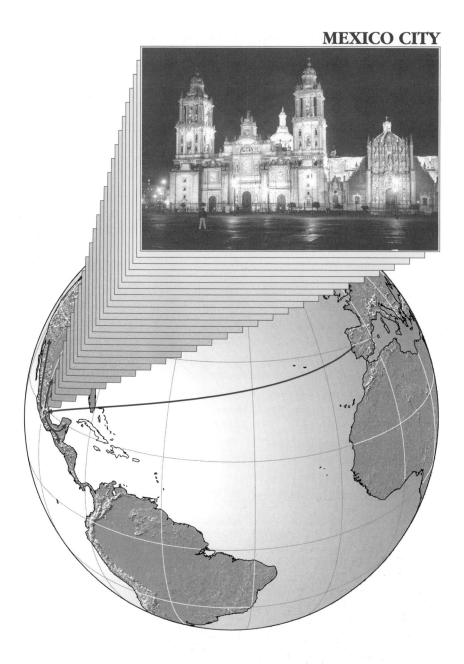

MEXICO CITY

FIGURE 3.4   *The changing interconnectedness of a city: from Tenochtitlán to Mexico City.* (a)   *On the left, Tenochtitlán (the Great Temple of the Aztecs by Eduardo Matos Moctezuma)*   (b)   *Above, Mexico City (cathedral on the Zocolo) Note how the geography of interconnections was extended and changed with the arrival of the Spanish as colonizers. The arrow-heads indicate the direction of the 'flow of power'*

107

Yet out of all this something new was created, something once again unique.
The contribution of the Aztec was not completely erased. If you go today as a
tourist to Mexico City, you will probably be most immediately conscious of the
Aztec presence through the heritage sites of the brochures: recent major
excavation works have exposed the foundations of the Templo Mayor which
once stood in the central precinct of Tenochtitlán (today it is on the corner
between the cathedral and the national palace); there is the immense Museum
of Anthropology in Chapultepec Park, with exhibits not only of Aztec culture
but of the enormous variety of other societies which lived in what we now
know as Mexico before the Spanish arrived; on the southern fringes of the city
there are the famous 'floating gardens' where (according to a tourist brochure
which I have at this moment to hand) 'flower-bedecked boats glide among
aquatic plants to the music of *mariachi* orchestras' but which are also the
remains of the *chinampas* where, before the Spanish arrived, people toiled in
the sun at their water-based agriculture.

Such tourist sites celebrating the Aztec 'presence', however, locate that presence
in the past. In fact it is more like an absence. There are other signs about the city
which bring its contribution closer. There are the place names – Azcapotzalco,
Tequesquinahuac, Tlalpantla Tenayuca, Nezahualcóyotl; there is the food – the
smell of cornbread in the morning (along with pollution) pervades the city;
there are certain understandings about the way of living everyday life. All this is
set in a city where the language is Spanish, the main religion Roman Catholic,

FIGURE 3.5   *Moctezuma meets Cortés (painted screen panel by Roberto Cueva
del Rio)*

and the dominant cultural references are to Europe and the USA. Out of this long history – indigenous and Spanish (and now independent modern Mexico) – has emerged, gradually and eventually, something new – something different from either of the two components. Mexico calls itself the mestizo state: the hybrid state. Mexico City is a mestizo city.

Yet to walk the streets of Mexico City is also to begin to understand the terms of this historical and geographical mixing. (And here we see that the indigenous 'presence' is not just in remnants from the past – it is an active part of the city today.) The three cultures of the square are here together. It is an unequal – and potentially and occasionally conflictual – mix. And this inequality is expressed in the spaces and the times of the city. The people who trip lightly in and out of the gleaming office buildings, perhaps rushing between 'urgent' meetings, have paler skins (look more 'European') than those who sit all day on the pavement outside, begging or selling embroidery. And the homes each retreat to at night, and the parts of the city where these homes are found, will be as different as it is possible to be. The dominant structures and images of Mexico City today revolve around those with paler skins, and are in the more 'modern'/ international mode. Indigenous lives receive more 'recognition' in the monument than in the economic and social structures of the modern city. The brightly embroidered clothing of the indigenous women on the pavement edge forms, in the terms of Chapter 2, an interruption from 'below the surface' into the dominant story of the city. The celebration of indigenous culture in *La Plaza* and in the Museum does not seem to extend to the streets of today.

1.3   REFLECTIONS

So what can be learned from this bit of a city's history? The most important point concerns a basic way of understanding cities in the world. This way of reading the history of Tenochtitlán–Mexico City emphasizes the city as a focus of wider networks of social connections. A city is a place where meetings happen: the meeting of the migrant with the already established resident; the meeting of traders in markets; the meetings of the powerful in dignified offices.

● Cities are set in wider geographical spaces within which they are in some sense a focus. That is the basic point.

But from that, other points can be drawn.

First, in general, cities are places of mixing. Precisely because of their role as foci *they bring together different histories*. That is what happened, on a grand scale, when the Spanish rode in to the Aztec city. But it happens, constantly and less dramatically, on a smaller scale all the time and in most cities. New people arrive, new stories are added to the ones already there. This process, however, is not a simple one and we can already note three things about it. To begin with, the different peoples, or histories, which a city brings together are themselves already mixtures. This may seem a strange point to make, but it is important. We tend very often to think of societies as having their places, of different cultures being based in particular areas. And we then interpret the historical increase in travel and communication, and the current frenzy of globalization, as linking together what were previously separate cultures and – on what is seen as the 'pessimistic' scenario – as breaking down all their authenticity and particularity and leading to a world in which 'everything looks the same'. This is not entirely untrue. Historically the amount and intensity of geographical movement and exchange has indeed increased enormously. But what is becoming much clearer from research in recent years is that the picture of once isolated societies gradually beginning to interact, or being invaded by the West, is a gross simplification. Thus Eric Wolf in his book *Europe and the People without History* (1982) has pointed to the enormous degree of interconnection between societies around the world in the centuries before the Europeans set out to explore. There were, he argues, no 'primitive isolates', no simply isolated cultures. And Janet Abu-Lughod in her book *Before European Hegemony* (1989) analysed what she calls 'the world system between 1250 and 1350' – and found in fact an intricate and active system of interconnections which formed a network of links (with cities as nodes): the 'world' which stretched from a fairly primitive Western Europe, through the Mediterranean, to India and China. When 'the Spanish' met 'the Aztecs' both were already complex products of hybrid histories.

Moreover, if cities are places of cultural mixing, then the social terms of that mixing will vary both historically and geographically. The different communities and cultures may remain relatively self-enclosed, inward-looking, or they may meld together, mix comfortably or irritably, or remain in conflict. There may be the 'ethnic quarters' of old Chicago, or the bounded and purified spaces described and deplored by Richard Sennett (Chapter 2, section 4.1), there may be the class and other separations produced by money and market, or there may be the glorious 'mixity' evoked by the likes of Jane Jacobs. And even this last will have its terms, its power relations. While there may on occasions be egalitarian jostling, recent more sceptical writing has pointed to middle-class invasions of inner-city areas as attempts to enjoy different peoples as exotica, not really to engage with difference but to treat it, rather, as 'local colour' (see, for example, May, 1996). The social terms of mixing in Mexico City – as we saw – are unequal and occasionally contested.

Finally on this point, whatever the terms of this mixing, something new will be produced. The coming together of histories in space will produce new stories. The long and painful difficulties of the Aztec/Spanish gave birth to the mestizo city. Putting this more generally, new 'geographical juxtapositions' produce new histories. We shall explore this important point further in the next chapter.

Second, geography (or 'space') may also be of fundamental importance in the expression and organization of this urban 'mixity'. As we have seen, in some cases there may be spatial integration (though the situations where this happens thoroughly are rare), but more often cultural variation will express itself, to one degree or another, in geographical terms. The enforced segregation of South African apartheid, the formalized enclaves for different nationalities in the treaty ports of China, the dual geographies of North African colonial cities – the tree-lined spaciousness of the French area against the tight huddle of the Arab medina; the sorting of cities by income group between suburbia and slum; the gay areas of San Francisco and Manchester. The word 'ghetto' originated in renaissance Venice where it described the areas in which Jewish people were forced to live: let out at dawn to do their business in the city, at nightfall they were obliged to return, the 'gates were locked, the shutters of its houses that looked outward closed; police patrolled the exterior' (Sennett, 1994, p.215). The important point here is that these divisions within the city are not just the result of mapping already existing, different communities on to distinct spaces. It is also that the spatial organization itself – the geography – is important in maintaining, maybe even in establishing, the difference itself. Chapter 2 (section 4.1) spoke briefly about this in its discussions of 'walls', and we shall return to the issue again. But once again the power relations vary. The Venetian ghetto was an expression of power relations and a hatred (and fear?) of people who were different. Here it was the dominated people who were walled in. In the gated communities of North American cities it is the relatively powerful, in an expression of their vulnerability, who wall *themselves* in. In San Francisco and Manchester the gay streets and villages are part of the process of establishing that community's existence – and right to existence. Geography and community and difference in the city are inextricably interrelated. One of the most troubling and continuing questions about urban areas is this: how is it best to organize spatially the different elements which go to make the variety which is so characteristic of so many cities?

Both of those first two points in a sense follow from the *third*: the fact, which we have emphasized in all the chapters of this book, that cities are essentially open, that they are places of wider interconnection. Moreover, and this is the really important point, these connections have to be constructed and actively maintained. When the Aztecs first arrived in the Valley of Mexico (they called themselves the Mexica then) they were considered by those already living there as a somewhat unsophisticated intrusion. It took years of negotiations and treaties, and defeats, before the city was even founded, and many years more before what began as a tentative settlement in the most unpropitious of locations had managed to establish itself as the centre of a network of trade and imperial tribute. The later connection to Spain was even more fiercely won: the weeks at sea, long miles across mountain and plain from the coast, to pull the orientation of the city towards Europe, towards Spain, to establish it as a focus in a new and different network of communications.

That process of constructing and re-constructing, negotiating and re-negotiating, their place in the world's geography is, as we shall see, crucial to the survival and success of cities. To drop out of networks can mean decline; to renegotiate your place within them can imply a change in trajectory, a different future. By the early nineteenth century the new Mexicans were beginning to find intolerable the imposed domination of their networks of contact, especially of trade, by the colonial power in Madrid. The battle for independence had as an important element the desire to re-orient Mexico yet again in the world, to re-balance and re-negotiate the geography of its networks of interconnection. And with national independence in 1821, the nature of Mexico City's placing in the world geography of interconnections changed once more.

When *La Plaza de las Tres Culturas* was created in the 1960s the very buildings expressed the significance that can be attached to being linked into wider networks. The contemporary buildings, set against the temple and the church, are – very deliberately – in the International Style of modernism. The choice of that style was itself an expression of what the Mexican government was aiming at in that period – industrialization, joining the 'modern' (for which read dominantly western) world. The very materiality of the built environment symbolized Mexico's intent to join in the networks of 'advanced industrial' countries.

ACTIVITY 3.2   Now, with that in mind – the significance (or at least the perceived significance) of a city's place in wider networks and the way in which that can be materialized in the built environment – turn again to the reading by Jane M. Jacobs at the end of the chapter. You have already looked through the piece in connection with Chapter 2. In your reading of it there, what was mainly drawn out was the way in which elements of the built environment, which already carry one particular heavy symbolism, can be re-worked into another. Their 'authority' is retained yet invested with a different and more contemporary significance. (Reflecting back on Mexico City's *Plaza*, it is possible to see how a similar process of resignification has gone on there too, with the pyramid and the church being now no longer only symbols of an authority and power manifested through religion, but also drawn into another scheme of representation of the cultural plurality of Mexico.)

Read through Reading 3A again now. Don't worry about the details – just appreciate them for the way in which they give such a clear indication of how even the apparently mundane minutiae of a planning inquiry can reflect much wider events and tensions.

As you read, do two things:

● make sure you are clear what changes were being responded to in this resignification of some of the City of London's most important buildings

- allow yourself to linger over certain phrases, especially in the early part of the chapter, which are trying to capture this business of the changing meaning of the built environment. Ones which caught my attention were

  'symbolic heart of Empire'

  'invested with meaning'

  'an active memory which inhabits the present'. ◆

The changes which lie behind the events discussed by Jacobs in Reading 3A are the shift away from the primary positioning of the City as the focus of the British Empire towards integration into a new global internationalism in which relations with Europe are of critical importance and within which London is no longer unquestionably at the centre. It is, precisely, *a reorientation of the City within a restructured wider geography.*

And that restructuring of the wider geography, of the networks within which the City is set, provokes effects within the City itself. Moreover, among those effects are what Jacobs calls 'active place-making events'. That is to say, the reconstruction of spaces and places within the City was an active part of the reordering of the wider relations within which the City is set and the aim was that the local reconstruction would respond to – and hopefully even influence – the remaking of the wider relations. 'Inside' and 'outside' the city are, then, intimately interrelated: the city is an open constellation of social relations around which it is hard to draw a line.

This complicated London planning saga began in the same decade that *La Plaza de las Tres Culturas* was being built. London and Mexico City – cities which had occupied very different positions in the building of the network of interconnections which, from the days of Hernán Cortés onwards, led into imperialism and colonialism. Cities at the opposite ends of empires. Perhaps this is reflected in these two 'active place-making events'? Mexico, in no doubt that it needs to get in on the networks represented by the International Style, while elements in London wish to assert a 'Britishness' both in nostalgia for a lost dominance and to cover a fear of rising competitors. And yet, finally, even the past is not 'local'. The Britishness (Englishness?) which some wish to assert is itself symbolic (just as are the church and the temple in Mexico) of a previous set of wider interrelations, in this case of empire and London's imperial power.

2 Megacities and global cities

2.1 SIZE AND POWER

Both Mexico City and London are now among the major cities of the world. What is more, between the founding of Mexico City in the place of Tenochtitlán and the development battles in London's financial quarter and the building of *La Plaza*, the role of cities in the world has changed beyond recognition.

At the turn of this millennium, and for the first time in human history, about half of the world's population is living, not just in 'cities', but in 'megacities'. That is to say, about 3,000 million people are living in urban areas each of which is home to many millions of people. In the late 1990s between 20 and 30 *million* people were leaving the countryside every year and moving into towns and cities. We have never faced a situation like this before. We are becoming, increasingly, an urbanized world.

The figures in Table 3.1 give some information about the world's biggest cities. Different sources give slightly different estimates. In part this is a result of the problem with which Chapter 1 had such trouble: how do you define a city? There are endless debates about how far 'out' the line should be drawn for statistics such as these. But the broad picture is the same.

TABLE 3.1 *World's largest urban agglomerations, 1992*

Rank	Agglomeration	Country	Population (millions)
1	Tokyo	Japan	25,772
2	São Paulo	Brazil	19,235
3	New York	United States of America	16,158
4	Mexico City	Mexico	15,276
5	Shanghai	China	14,053
6	Bombay	India	13,322
7	Los Angeles	United States of America	11,853
8	Buenos Aires	Argentina	11,753
9	Seoul	Republic of Korea	11,589
10	Beijing	China	11,433
11	Rio de Janeiro	Brazil	11,257
12	Calcutta	India	11,106
13	Osaka	Japan	10,535

Source: Castells, 1996, p.404

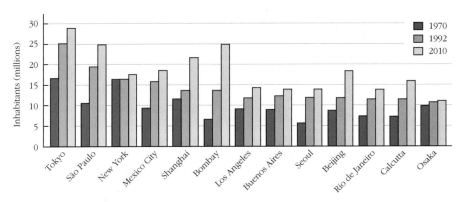

FIGURE 3.6 *The world's largest urban agglomerations (more than 10 million inhabitants in 1992) (Source: Castells, 1996, p.405)*

ACTIVITY 3.3 Look at Figure 3.6 and note down which will be the top ten cities, by size, in 2010. These are the world's megacities. Look at them and see what strikes you about the cities in this list. ◆

Perhaps the most obvious point is that the majority of the world's biggest cities are now – and even more are going to be in 2010 – in what is termed the 'Third World'. There is not a single European city in Figure 3.6, for instance. Another issue which sprang to my mind is to do with importance or influence.

ACTIVTY 3.4 If you were to draw up a list of the most *powerful* cities in the world, which cities might be in it? Make a list of, say, 5 or 6 cities. If you have any doubts or thoughts about this list, jot them down too. Now compare this list with the list of the biggest cities. How do they compare? ◆

It seems that, so far as cities are concerned, there may not be a direct and simple relation between size and power. São Paulo is not more powerful, in my view, than New York, for instance, though it is bigger; nor is Bombay (Mumbai) more powerful than Los Angeles. And I would have thought that London, certainly in some terms, such as finance, was more powerful than all but two of the cities listed, yet it does not appear. Some cities can be huge, yet not very powerful players on the world stage. And vice versa – think of Washington, DC.

That, however, raises the question: what does it mean to talk of a city as powerful? (This was the note I jotted down on my list: obviously, one's list will depend on how one defines 'powerful'.) We can draw on the discussion in the last section to explore this. There it was argued that one of the key things about cities is that they are the foci of wider geographical networks of social relations. Moreover, as we also just began to see, those social relations will be the bearers of various kinds of social power. The Aztecs made Tenochtitlán a dominant focus in the hierarchical geography of social relations in mesoamerica. In its own space, at that time, the city was the single most important locus of power. But in its transformation from Tenochtitlán to Mexico City, not only did it experience a complete reorientation in the network within which it was embedded but it also found itself, although still dominant (now in different ways) within its region, subordinated to an even greater centre of the relations of power – this time in Madrid. Jacobs (1996) is quite explicit about the connection between a city's power and the significance and nature of its interrelations. She writes (in Reading 3A), for instance, of the 'City of power' that it was 'a symbolic site of a Britain made great by its global reach'.

So, on this basis, we can suggest that a city's power can be assessed by the degree to which it is the source of social relations of power which radiate from it. It is this kind of definition which has given rise to the concept of the *global city*. Indeed you have already met this concept briefly in the reading on London.

One of the most important theoreticians of the city who has worked on this concept is Saskia Sassen. In her book *The Global City,* she writes the following:

> ... the combination of spatial dispersal and global integration has created a new strategic role for major cities. Beyond their long history as centers for international trade and banking, these cities now function in four new ways: first, as highly concentrated command points in the organization of the world economy; second, as key locations for finance and for specialized service firms, which have replaced manufacturing as the leading economic sectors; third, as sites of production, including the production of innovations, in these leading industries; and fourth, as markets for the products and innovations produced. These changes in the functioning of cities have had a massive impact upon both international economic activity and urban form: Cities concentrate control over vast resources, while finance and specialized service industries have restructured the urban social and economic order. Thus a new type of city has appeared. It is the global city. Leading examples now are New York, London, and Tokyo.

(Sassen, 1991, pp.3–4)

This passage is doing two things. First, it gives hints as to how we might think about the power of cities in a general sense. Second, it talks about the specific form of power which is dominant today and which is concentrated in global cities.

On the first, it is clear that a city's power is defined here in terms of its relationship to wider areas. Powerful cities are 'command points', they 'concentrate control', and they do so through their nets of interconnections. There is no mention here of the size of a city's population.

On the second issue, Sassen is making an argument about what form of power is dominant today; she argues that it is economic power, and most especially that which is embodied in banking, finance and specialized services. She is, in other words, talking about one kind of network of social relations, presenting it as currently the most powerful, and concluding that the cities which are the crucial command points in those networks are today's 'global cities'. These cities, in very important senses, 'run the world'. (See also Allen and Hamnett, eds, 1995, especially Ch.3.)

However, while the global networks of banking, finance and related services are the most powerful in the world today, there are three other considerations we should bear in mind in our own exploration of cities. *First,* that banking and finance and so forth have not always been the most powerful connections between cities. *Second,* that there are still today many other forms of power with their own networks and their own hierarchies and distributions of control or influence. And *third,* moreover, that not all networks need take the form of hierarchies with command-centres at the top. There could be networks which are flatter, or more egalitarian, or which function more simply as channels of equal exchange, communication and facilitation.

To take one completely different example, religious centres have in the past been the foci of powerful geographies (one thinks perhaps of Rome and the global power and influence of the Vatican), and they remain significant (though in different – and arguably less 'powerful' – ways than banking and finance) today. As well as Rome, we could point to Mecca and Jerusalem. Or, again, if one were thinking of the networks of global culture, Los Angeles (Hollywood), Paris and London might figure prominently. And if one broke that down into more specialized cultural networks the picture might change again: San Francisco and Sydney for the gay and lesbian networks, perhaps; Bombay at the centre of the cultural networks maintained by the diasporic communities from much of India. Or there can also be what one might call 'networks of ideas'. The construction of the 'modern' buildings in Mexico City's *Plaza* did not only reflect the spread of particular dominant cultural styles (the International Style); it also embodied the spread of ideas about what was necessary to *become* 'modern' – what kind of economic development strategies, for instance (in this case, industrialization). This kind of spread of dominant ideas (for instance about economic development), and of particular dominant theories (theories about cities, for instance), is an element in globalization. It may be important, for survival, to sign up to them; their very dominance may make more difficult the imagination and pursuit of alternative futures.

So far, then, we can conclude the following:

- there are numerous networks of power and influence, and they do not simply map on to one another
- these networks are of differing significance: they differ in terms of the kinds of social power which they carry as well as in the numbers influenced and in the degree and nature of that influence

- individual cities have different balances of all of these, and different positions within them

- cities are foci of such networks, but they are so in very different ways, giving them distinct kinds and ranges of influence. And, moreover, the position of any city within such networks may change over time.

We are beginning to put flesh on the rather abstract proposition set out at the end of section 1.1: that cities are nodes of social relations in time and space.

2.2   IMPACTS AND FUTURES

Let us take this notion a bit further. If cities are indeed intense foci of social relations in time and space, then

- their impact and effect will stretch way beyond their physical extent: in the world of today, with cities so dominant, the question of what happens to the world beyond them becomes increasingly urgent

- changes in a city's place within these networks can deeply affect its fortunes and its character.

2.2.1   What of cities' impact on the non-urban world?

Here we can only touch on this question, through one example: these cities have to be fed. In general terms, of course, they have to be supplied with a wide range of goods and services, much of which is produced within them. But food poses particular questions. Very little primary production of food takes place within cities. Remember Chapter 1, which spoke of 'the city that does not know how it is fed'? Indeed, the networks of trade and production, the vast areas and numbers of people, which are engaged in feeding the cities of the world, are probably the least acknowledged ways in which cities function as nodes within the organization of economic life on the planet.

Until only relatively recently even major cities in the First World would draw a high proportion of their food requirements from quite a local area. Very often such cities, where the land was appropriate, would be surrounded by mile upon mile of market gardens, sending fruit and vegetables (the classic 'perishable goods') to the urban population. Other foodstuffs would be brought in from farther afield (remember Chicago) and a few 'exotic' things (from pepper and spices in the early days to bananas more recently) would be shipped in over vast distances. For large, rich, and sophisticated cities that picture has now changed utterly. Every day the whole of the planet is raided for an incredible range of goods: green beans from East Africa, kiwi fruit from the antipodes, avocados from Latin America, strawberries available whatever the season. A whole range of factors has contributed to this, including the growing relative wealth of (the majority of) the populations of First World cities – a population now able to demand not only exotic goods from all over the world but also all particular foodstuffs all the year round. A further factor which has been important is the changing international organization of agriculture. No longer do our green beans come from a smallish plot in a market garden not too many miles away; rather they are flown in from another continent where whole villages are recruited into their production. The lives of people thousands of miles away are utterly transformed.

This is an example of the change in what is sometimes called the city's *footprint*. Although the term calls to mind the image of a solid area – literally a stamp upon the Earth – in fact a footprint can better be imagined as yet another example of the complex networks of relations which spread out from and focus in upon the cities of the world. Networks of trade, the massive production links of food multinationals, the world bodies and financial markets which deal in food, and so forth. A city's footprint refers to the geographical spread of its impact upon the world. The tentacular reach of First World cities is vast.

Such global networks of food production for cities pose problems and raise issues. They raise again in ever sharper form the question of the relation between town and country (but now they raise it on a world scale) and even of the relation between 'Society' and 'Nature'. They also raise questions of sustainability. There is, of course, the question of whether the cities can continue to be fed. But there are also two other issues. On the one hand, there is the issue of the sustainability of the Third World *rural* areas in which so much of our food is now produced. What does the current degree and rate of urbanization mean for the world's non-urbanized areas and people? How does reorienting the economies of these areas into this set of global connections affect the lives and livelihoods of the people in them?

And, on the other hand, there is the increasingly pressing issue of transport and pollution. What must be the fuel costs and environmental impacts of the transport of food around the world, followed by its distribution within cities and, then, as the final *coup de grâce*, the fact that people travel miles in their cars to out-of-town shopping centres to buy it and bring it home? In Britain the average distance travelled by items of food rose by 50 per cent between 1978 and 1993 and that *despite* the fact that the amount of food being transported remained static at around 300 million tonnes a year (Lang, 1997). (Over one-third of the growth in the entire national road freight in the same period is attributable to food, drink and tobacco – more than for any other major commodity group.) Such facts might lead us to question the viability and the acceptability of some of these 'nets of interconnections'.

And yet I have been writing here of 'rich and sophisticated' First World cities in 'the North' of the planet. The same pattern does not hold for all the cities of the world. In cities of the world's South there may well be an equally rich and sophisticated elite whose habits of consumption are very like those of the First World elite. But much of the population of Third World cities lives nothing like this. The resources they draw on are small and the spatial range on which they depend is far more constricted. Even for some quite large cities, in Africa for instance, the sets of interconnections through which the bulk of the city is largely fed have remained, at least until very recently, geographically relatively local. Their footprint upon the Earth is comparatively small and light.

2.2.2 Changing networks and changing fortunes

Having grasped the idea of the complex of networks within which cities exist, we need also to imagine those networks changing over time. We have already met this issue: the question now is the relation between changes in their interconnectedness and changes in the fortunes of cities.

Most dramatically, cities can become disconnected from the networks which previously sustained them. Tenochtitlán–Mexico City was detached by the policy of the Spanish from its previous imperial connections. Its insertion into a new network allowed its re-establishment in a new form, and its expansion of that network on independence ensured its continued growth. But, sometimes, issues of disconnection can be more complex, and patterns of disconnection quite selective.

For one thing, such 'disconnection' can happen by design, by the implementation in some way of a definite strategy or policy for the city. There have been many examples through history, though usually related to national rather than specifically urban strategies. There has, for instance, been an attempt by certain Islamic countries in recent years to design and build a future which does not follow in every detail the model of development exemplified by 'the West'. That strategy has meant breaking off certain relations. Certain films have been banned, and some kinds of music; the arrival of western cultural influences has been carefully monitored and controlled. This meant voluntarily opting out of some of the networks – particularly cultural but also economic – which were in principle available. For the character of the networks in which a city (or country) is involved influences not only the degree but also the nature of its growth and development.

This leads to an important consideration. For although the 'power' of a city may be defined (as earlier in this chapter) in terms of its involvement in wider networks, and its position of control or influence within them, we might want to make a distinction between 'power' and 'success'. We have already made a distinction between power and size. Now we must hold separate – at least for the moment, to question them further – this other pair of concepts. Is a powerful city necessarily the same as a 'good city' in which to live? And by whose lights? For what is clear from the Islamic strategy just mentioned is that there was a willingness to forgo certain kinds of power and influence on the global (western) scene in order to try to develop a kind of society (and city) of which they more approved. In that context a different *kind* of city would have been rated a greater success. Now, it can also be argued that what was really at issue was a feeling that, as matters currently stood, Islamic cities and countries would anyway be subordinate within a western-dominated global system and that greater power might be gained by constructing alternative networks. But, again, the notion of power, and of its relation to 'success', is further problematized. What is at issue is not simply a city's degree of openness or interconnection, but its influence and control within these relations. Cities which deliberately cut themselves off from certain networks will often have been subordinated within them – the aim is generally to protect themselves from incoming influences.

Some places, of course, are cut off by explicit strategies of exclusion on the part of others. The boycott of Cuba by the USA forced Havana to join the network of interconnections centred on Moscow. The boycott of South Africa, because of its policy of apartheid, altered the trajectories of Cape Town and Johannesburg. (And, of course, dropping out of one set of networks does not mean leaving them all: a city which is not part of global finance may still be open to flows of migration, for instance.) But most cities slip out of networks as part and parcel of economic change and decline. When the border was drawn between East and West, dividing Germany in two, the city of Hamburg lost the bulk of its hinterland and the immediate effect was a diminution of economic activity. Or from a completely different period, Abu-Lughod (1989) describes how the fair-towns of Champagne in France, which had flourished greatly in the thirteenth century, were plunged into decline when, among other things, new trade-routes passed them by completely.

More frequently, however, it is economic change and decline itself which provokes a shift in, or reduces, a city's interconnectedness. For cities are also major producers of goods and services. The decline of Glasgow's shipbuilding and port industries changed dramatically the way in which that city was economically linked into the rest of the world. Old connections were lost, or became attenuated. It was a decline which reinforced the city's loss of trade as the orientation of the UK as a whole slowly moved around, from the Atlantic towards continental Europe.

All this raises a significant point about how we think about cities. In this chapter, the stress has been laid on cities as foci of wider interconnections. And this relationship between interconnectedness and survival and growth was also made in Chapter 1. But, as that chapter also pointed out, simple interconnectedness is not enough. It also matters what happens *within* cities, what is *made* of the interconnections. The mere criss-crossing of wider networks of social relations (their simple intersection) is not enough to produce a city. At best it might result (using an example of economic relations) in a staging post, a transhipment point, a locus of simple exchange. For development towards city-dom what is needed is positively activated inter*action*. This could mean the bubbling-up of new activities, it could mean specific policies to trade on or maintain the potential effects of intersection (to turn it into interaction). In the history of early Chicago some of its people spent much time and effort in stabilizing the flows which went through the city and in doing something with them while they were there.

So, to pick up again that abstract definition – the city as a node of social relations in time and space – we can now see that such 'nodes' must be more than simple intersections, that they must be activated to produce additional effects (relations and activities) to turn them into the *intensities* which we know as cities.

However, cities are – as this chapter keeps on stressing – also essentially *open;* they are foci of interconnections with a wider world. Cities are *open intensities.* And both aspects of this interpretation of them are important:

- on the one hand, there are the new activities and initiatives, the new stories (conflictual or co-operative), the new mixtures, which can be actively produced out of the geographical juxtapositions within cities
- on the other hand, much of the energy of a city comes from its openness to outside: from the new migrants who arrive to add to the mixture and maybe participate in the creation of new cultures, from trading connections and cultural influences, from flows of money.

If a city which did nothing with the intersections on which it was based would be likely to stagnate, so would the city which completely walled itself in.

All over the world and year after year we hear of strategies for cities to revitalize themselves through reasserting or increasing their influence on 'the world-stage' (or, rather, on one of the many such stages). Glasgow set itself up as a European 'city of culture', and was internationally confirmed as such through competition. Every few years cities from around the world compete to hold the Olympic games, or the winter games, or some other sporting event. Virtually every city, in the West at least, runs an international advertising campaign to attract to itself tourists, or investment, or some other kinds of business, from around the world. What is going on here is one element in a global competition between cities for survival and growth. Each is trying to climb the hierarchical network of which it is already a part or which it deems most appropriate for its future development: the 'high culture' network for Glasgow and Barcelona; the financial network for Singapore. Each is trying to re-negotiate its position within geographical networks of interrelations in the hope that by doing something with those more powerful interconnections it will be able to change its future trajectory.

But there is one further point here, which you may already have thought of yourself. This is that 'cities' in themselves are not actors. 'Cities' as singular entities do not really design strategies for 'themselves'. In Chapter 2, there was frequent reference to how attempts to produce singular images of cities were in fact repressing, covering over the existence of, other stories, maybe dissident voices. If there is one thing we can say already about cities, it is that they are very plural places. And that means that different interests are involved in different strategies, and that over most strategies there is at least some degree of dispute. Almost every time a city council decides to submit a bid in an international competition to host some major event (the Olympics, say) there are groups within that city who argue that this is not the right way forward. Glasgow's 'city of culture' strategy provoked enormous controversy. There is an intermittent and fierce debate in London about how much priority to give to 'the City'. It may do wonders for some, and for the Square Mile and the Docklands, and certainly it increases 'London's' power in the world. There is no doubt that the power and importance of London, in the global arena, is based at least in part on the size and vitality of its banking and finance. But others ask what this power does for the majority of Londoners, and whether it makes for a 'successful' city.

So we might want – in thinking about strategies for cities – to make a distinction between success and power on the world stage. And we might also want, in attaching any of these characteristics to a city, to ask 'for whom?' We might *then* want to turn back to that phrase in section 2.1 – 'what does it mean to talk of a city as powerful?' – and to modify it. 'A city' being powerful may well mean only for some of the inhabitants within it. We may want to ask also, what of the rest?

2.3   OTHER APPROACHES

We shall be returning to some of these big questions in later sections and in the next chapter. But for the moment let us step back and draw out some elements of where we have got to. In particular, how are we approaching the issue of 'cities in the world'?

There are two common approaches to the issue which are explicitly rejected in this book (and in the series of books of which this is but one). The first approach is to tell a single story of 'the development of the city', a story which classically begins – after a few nods in the direction of 'prehistory' – with Athens. The Greek city-state is where we start so many of our histories of western society and it might seem to be especially appropriate for cities. After all, so many of our urban words – *polis*, for example, as in metropolis etc. – derive from that time and that place. From Athens the story will then proceed through a series of cities which are taken as exemplars of an age: through Rome, Venice, Amsterdam, Paris, London and New York, typically ending up with Los Angeles, famed as the quintessential 'postmodern' metropolis. We reject that approach for two, linked, reasons.

ACTIVITY 3.5   Why do you think there might be difficulties with an approach to 'cities' which deals with them as a story which runs from Athens to Los

Angeles? Think about what you have read in the chapter so far: from that, you might be able to come up with at least one criticism. ◆

Our main criticism of this approach is that it tends to include in its history primarily, if not entirely, cities of the West, of today's First World. Yet, as we have already seen, not only is the imminent future of urbanization – in terms of the size of the phenomenon – now much more importantly a Third World issue, but also many huge cities lie quite outside that linear history of Athens to Los Angeles. Where, for instance, would Tenochtitlán fit in, or the cities of thirteenth-century Eurasia discussed by Abu-Lughod – Samarkand, Constantinople, Baghdad and Cairo; Calicut, Malacca and Canton? And that relates to our second reservation about this approach: that it sees urban history as one single line of development. Rather, we would argue that, although there clearly are many commonalities, different cities do have different histories, and can be expected to have different trajectories in the future. Thus, we do not see Calcutta, for example, as in any simple way 'following' Los Angeles.

The second approach about which we have our doubts relates to this. This is the approach which talks in terms of a single world system and a single global hierarchy of cities. Instead of a single history, as in the first approach, here we have a single, integrated geographical system. The arguments just presented also militate against adopting this framework. Rather – we would argue – there is a multiplicity of 'systems', of interlinkages and interconnections each with their own shapes and contours, hierarchies and rankings. Any one city may play a role in one or more of these. Within that complexity – within that 'power geometry' – different cities have their own trajectories and there is a constant process of the making and unmaking of connections.

3 *The city without and within*

3.1 DIVIDED CITIES

Just as there is no single system of cities, so also it can be problematical to think in terms of a single entity – 'the city'. Here the notions of cities as locations of mixing, interaction and differentiation and of cities as nodes in networks come together. As was argued in the last section, cities are plural places, difficult to grasp as singular entities with one voice. We can now take this further: if not all cities are included in all networks, neither are 'whole' cities simply included or excluded. While 'Birmingham', say, might be closely tied into (though not dominant within) the global geographies of multinational manufacturing production, it is also of significance on the international circuit of classical orchestral music, and is an important centre in the networks which link together the internationalized communities of migrants from India and Pakistan. Within the city, some people may tie in to all those aspects of 'Birmingham', others may live lives which connect with some but not all, and yet others might feel themselves excluded from more or less every one of those ways in which 'their city' is a focus of international significance.

As, over history, the nets of interconnections within which cities are embedded are reworked, as they develop, are altered and adjusted, so too the place of cities and parts of cities within them is shifted. New juxtapositions and new mixings of social groups and activities are produced. But also, so are new divides. These hierarchies of power and influence do not extend to everywhere; exclusions can be produced in the very process of the establishment of new patterns of interconnection.

Let us look at a particular case. Perhaps one of the most significant and far-reaching 'reworkings of interconnections' which has been going on in recent years is that associated with economic globalization. It was an aspect of this phenomenon that Sassen was alluding to in the quotation cited earlier – the establishing of world-spanning networks of banking, finance and related services and the focusing of these networks into a number of hierarchies centred on particular cities. This is one of the central – and most often cited – characteristics of today's economic globalization, which would also include the increasing internationalization of other economic sectors – the production of cars, for instance, and the production and trade of agricultural produce. These major reworkings of the networks of world economic geography have drawn new areas in to global connections and, just as actively, have excluded others.

For example, one element of globalization over recent years has been the internationalization of production and trade in agricultural produce. Increasingly, this is in the hands of a small number of multinational companies and increasingly, too, Third World countries have been encouraged to specialize in growing food crops for export to richer areas of the world. This has been combined with an emphasis on 'free trade'. (It is also part of a wider political strategy of neo-liberalism: see **Allen, Massey and Pryke, eds, 1999**.) We already touched on this briefly in the discussion of 'footprints' in section 2.2. There we were concerned mainly with the impact of the need to feed the richer First World cities, and of the way in which this was being done. Here we find that one of these impacts has been precisely on the cities of the Third World. For one result of these changing demands and networks of production and trade has been the massive migration of people from Third World countryside to Third World cities. No longer able to compete with imported mass-produced foods from elsewhere millions of peasant farmers have left the land (Lang and Hines, 1993). It is this (combined with a crackdown on immigration into most First World countries) which is one factor behind the much faster rates of growth of cities in the Third World (see Figure 3.5). In other words, this growth has been produced, at least in some measure, as a result of the reorganization, and very importantly the extension, the geographical *stretching*, of particular networks of power in the world economy.

Yet when they arrive in the city (the focus of other globalized relations) these migrants may often find themselves excluded there too. They are, in a very real sense, *dis*connected. Manuel Castells, who in his book *The Rise of the Network Society* has produced an enormous amount of information and analysis of this phenomenon, writes as follows (and note that here he is including 'global cities' within his category of 'megacity'):

> Megacities articulate the global economy, link up the informational networks, and concentrate the world's power. But they are also the depositories of all those segments of the population who fight to survive, as well as of those groups who want to make visible their dereliction, so that they will not die ignored in areas bypassed by communication networks. Megacities concentrate the best and the worst, from the innovators and the powers that be to their structurally irrelevant people, ready to sell their irrelevance or to make 'the others' pay for it. Yet what is most significant about megacities is that they are connected externally to global networks and to segments of their own countries, while internally disconnecting local populations that are either functionally unnecessary or socially disruptive. I argue that this is true of New York as well as of Mexico or Jakarta. *It is this distinctive feature of being globally connected and locally disconnected, physically and socially, that makes megacities a new urban form.* A form that is characterized by the functional linkages it establishes across vast expanses of territory, yet with a great deal of discontinuity in land use

patterns. Megacities' functional and social hierarchies are spatially blurred and mixed, organized in retrenched encampments, and unevenly patched by unexpected pockets of un-desirable uses. Megacities are discontinuous constellations of spatial fragments, functional pieces, and social segments.

(Castells, 1996, pp.405–7)

This is not a minor phenomenon. It has been estimated that at the turn of this millennium 60 per cent of the population of Asian megacities lives in squatter settlements. In city after city the population is divided by great chasms, dividing élite from poorer areas. Nor is this a phenomenon exclusive to the Third World. The image of shining office-blocks towering over inner-city poverty can be drawn from any part of the globe. As Castells argues, '… this is true of New York as well as of Mexico or Jakarta'. Different parts of a city may have different connections to the wider world, be linked in to different sets of networks with different degrees of power, and in some cases find themselves excluded. There are some terms which Castells coins in the course of his analysis which give one pause for thought about the nature of the current development of cities, and about their futures. The term 'structurally irrelevant people' is one such. What he means is simply those excluded from the kinds of networks of social relations – the kinds of power geometries – which link together and are the basis for the power of major cities. But where such exclusion is from *all* significant forms of power, the implications – both social and political – are likely to be dire.

3.2   PROBLEMS IN THE CITY

ACTIVITY 3.6   Now study the extracts about Bombay and Los Angeles, by Rahul Mehrotra and Camilo José Vergara respectively. These readings put flesh on the bones of the last section and also take the argument a little further. ◆

EXTRACT 3.1
Rahul Mehrotra: 'One space, two worlds: on Bombay'

In Bombay, the rich and the poor live in distinct physical environments and locations, one static, monumental, and on the high areas of the city, the other sprawling along the transport lines and into any available interstices or crevices.

Poor rural migrants, in huge numbers, are shaping the culture and form of the city – today it is like a bazaar, a kaleidoscope of snapshots and symbols overlapping to create an often incomprehensible mosaic. In fact, the bazaar, a chaotic marketplace of shops, stalls, and hawkers, can be seen as the metaphor for the physical state of the contemporary Indian city

– an informal enterprise zone expressing energy, optimism, and a will to survive outside any formal system.

In the 1960s, the bazaar swept across Bombay, sprawling along transport lines, on slopes, underutilized land, undefined or unpoliced pavements, and on any other interstitial space it could find and occupy. In the process, it blurred beyond recognition the physical segregation inherent in the colonial city structure that had survived until then. The bazaar completely altered the exclusivity of the two domains, the Indian town (in the north) and the Western city (to the south). Emblematic of this physical coalescence of two worlds in one space are the bazaars that occupy the Victorian arcades in the part of Bombay that formed the core of the British town …

Source: Mehrotra, 1997, p.40

EXTRACT 3.2
Camilo José Vergara: 'One face of the future: Skid Row, Los Angeles'

'There is nothing like this in Russia.'
(Ludmilla, manager of Prime Fish Co., Skid Row, 1996)

'Do not look at the homeless, look at these urban industries.'
(Historian Robert Fishman, pointing out the positive aspects of Skid Row, 1996)

Making toys, furniture, and artificial flowers, storing and processing fish, and serving the homeless are the main activities of L.A.'s Skid Row. As one learns more about the individual complexity of each of these businesses, their interdependence, and their relationship to the missions and the street people, the more extraordinary this urban economy seems, thriving as it does among misery, disease, and despair.

The missions came to the area first. The Fred Jordan Mission is over forty years old, and the Union Rescue Mission, now in a new building, has been here for ninety years. Many of the hotels built long ago to accommodate travelers became SROs (single-room-occupancy hotels) after the train station was relocated, providing another cheap place to live; some offer a bed for as little as fourteen dollars a night. The missions were followed by soup kitchens, health facilities, and methadone clinics. People know to drive to this part of the city to bring food and clothing for the homeless, and their charity gives further impetus for the down-and-out to congregate here.

The worlds of the destitute and of local industries interact. Businesses hire homeless people to load and unload trucks, to sweep the streets in front of their facilities, and even to discourage other street people from loitering and scaring away customers.

The complexity of the large missions is awesome. A variety of populations use the buildings: schizophrenics, manic depressives, drug addicts, unwed mothers, families displaced by fire, volunteers, and others. Accommodating such diverse populations without friction requires that the different categories of people who share the same building have only minimal contact with one another. Sections of floors are locked, and key cards and number codes are required to go from one section to another. Guards sit at checkpoints, directing the flow of people. The floors are made of concrete for easy cleaning, which helps deter the spread of disease. Even though a visitor may see few residents, their presence can be sensed on the other side of walls and on other floors.

From the outside the new missions look large and expensive. Their exteriors are warm, welcoming, even playful. Everything denies the strict regimentation and despair found inside. A local minister described the new missions as 'partly a lie,' comparing them to 'a strange woman that has all kinds of diseases and puts on make-up to cover them.' A critic commented, 'we would like to make the homeless invisible.' Another replied by asking, 'What is the alternative?'

The seafood processed in the local plants include salmon, shrimp, and other expensive varieties. One of the facilities manufactures a choice type of beef jerky, a delicacy for the Japanese. The businesses make money – and the architects who design the missions have their work discussed and featured in the media. Those with a chair or a bed inside a mission and the homeless in the streets make do the best they can. In its capacity to aggregate misery and profit in a large urban space, Skid Row shows one face of the future.

Source: Vergara, 1997, p.11

First, what we see here is the city as a 'space' in which different stories meet up. This is not passive poverty within a thriving city: the bazaar and Skid Row are active worlds of their own. The Bombay article talks of 'energy' and 'optimism' and the inhabitants of Skid Row are tied into an urban economy which is 'thriving', even if it is 'among misery, disease, and despair'. We might, therefore, elaborate a little that quotation from Castells. Today's urban world is not made up of only one big global story (such as the story of finance and banking) from which others are excluded. There are other stories too. The rhythms of the bazaar may not be those of the wheelers and dealers of the financial global economy, but they are just as vital (though the paces of life of those in the Los Angeles missions, where so many different individual stories are brought together, must be incredibly varied – from the hyperactive to those who sit and stare). There really are different trajectories coming into close proximity in these places.

Moreover, one of the things Castells stresses, elsewhere in his books, is the greater degree of global interconnectedness of the rich and powerful. He is certainly correct in this. We must, however, be aware too of the complexity of the interconnections even of those who are excluded from the dominant networks. They are not simply 'local' people, counterposed to the globality of the powerful.

Imagine the bazaar ...

- maybe some of the people there have come from the countryside, squeezed out of their former agricultural lives by the arrival of multinational producers (see sections 2.2 and 3.1 – their lives, then, have already been touched by globalization, though here they have very much been 'on the receiving end')

- they arrive in the city, but no way can they find employment in the formal economy (they are excluded from those links into globalization – they are among the 'structurally irrelevant' people of Castells' account)

- however, here in the bazaar they buy and sell goods from near and far: from the small businesses of the city, from the surrounding countryside, from 'the West' – western cigarettes, watches, trinkets – and many things which mimic those of the West (here they are constructing their own economic interconnections – fragile and tenuous maybe, but nonetheless quite extensive)

- and they have family: parents maybe still in the village, a son and daughter-in-law who are now in England ...

In these divided cities the different stories which are lived out side by side are certainly grotesquely unequal in wealth, health and quality of life in material terms. But they are all the product of complicated interweavings of networks of social relations. This, then, is our next point.

Second, both these sets of writing and photographs clearly present pictures of dislocation and disconnection within the city. In Bombay's 'bazaar' there is a will to survive 'outside any formal system'. Certainly the people involved in this 'chaotic marketplace of shops, stalls, and hawkers' are not themselves directly connected into powerful global networks. Yet neither is there complete isolation (there are no 'isolates', in Wolf's terms – see section 1.3 above) – no really isolated cultures. On Skid Row, 'The worlds of the destitute and of local industries interact': the people of this 'disconnected' area in fact often provide a labour force for businesses producing goods which are sold in the richer parts of the city. (Remember the quotation about Cerrada del Cóndor in Chapter 2, page 72?) And there are connections too through the charities. Connections, then, but what needs to be analysed is the social form of those connections: what is evident is that the place of Skid Row and the bazaar is marginal to networks of what we might think of as real power, certainly of economic power.

Let us pause for a moment and draw out some of the many connections into which these lives are inserted and through which they are lived. As we just said: what we need always to bear in mind is the social form of these connections, and who has power within them.

Imagine the Los Angeles hostel ...

and the different kinds of spatial organization tying different worlds together, and holding them apart:

- the busy 'complexity' and 'interdependence' of the businesses, the missions and the street people within this local urban economy, including both trading links and employment

- the visits from individual people, driving down in cars from wealthier areas, bringing charitable donations (and so there is a further concentration of the needy)

- the fierce divisions of space within the missions themselves, so that they can 'accommodate such diverse populations without friction' (think back here to other spatial separations you have read about in this book)

- the startling difference between the welcoming appearance of the outside of the missions and the sense of despair which envelopes you as you step inside (what is this about?)

- an architect on a podium in an expensively furnished conference hall in a city miles away; the lights are dimmed and the architect, giving a presentation of recent work, turns to the large screen, upon which appears a picture of the mission.

Third, what these extracts most powerfully give evidence of are the huge divisions which so often split apart the populations of today's cities. For all the excitement, attraction and growth in cities they also hold out the converse of these things: spiralling inequality, a terror every now and then that things might get 'out of control'. This notion that the chasms of inequality might lead to social unrest comes out a number of times in Castells' book. Have a look again at the quotation. He writes of 'those groups who want to make visible their dereliction, so that they will not die ignored in areas bypassed by communication networks', 'people ready to … make "the others" pay for it' and of 'retrenched encampments'.

So social explosion is one challenge facing twenty-first-century urbanization. But it is by no means the only one.

ACTIVITY 3.7   Now read Extract 3.3, taken from a World Bank report by M. Ismail Serageldin. ◆

EXTRACT 3.3
M. Ismail Serageldin: 'A decent life'

Since the end of World War II, the earth's population has increased from two billion to five-and-a-half billion. No matter what is done by population programs now, one billion more people will be born by the end of the next decade, and one billion more in the decade after that. Only in the third decade from now can one begin to predict a slow-down in population growth. Moreover, practically all growth will occur in the developing world, where incomes are a fraction of those in industrialized countries. The urban population of developing countries has grown from fewer than 300 million in 1950 to more than 1.7 billion today, and it is projected to more than double in the next twenty-five years, reaching nearly four billion. In the 21st century over 50% of the global population will be urban.

Cities are also growing in size and complexity. In 1960 only 100 cities had more than one million inhabitants. Now there [are] almost 400 such cities, and over the next decades there may be as many as 650. Attention tends to focus on megacities such as Mexico City and Shanghai, but population growth has affected cities and towns of all sizes; secondary cities such as Kano, Nigeria, Surabaya, Indonesia, and Guadalajara, Mexico, have become metropolises of two to three million in the last decade and continue to grow rapidly.

Cities in developing countries have much larger populations than cities in developed countries; thus the scale of their urban expansion will be unprecedented. For example, over the next forty years, the increase in the urban population of India will be more than double the population of France, Germany, and the United Kingdom *combined*. The management challenges posed by growth on this scale will be substantial.

Developing countries also face the challenge of rapid urbanization with far fewer resources than the developed countries possess. The per capita income of the United States was 50% higher than that of Kenya and India at an equivalent level of urbanization. Added to this difficulty is the fact that as the global economy becomes increasingly integrated, cities in developing countries are under great pressure to compete. Many of these cities are now organized very inefficiently and will need to make major adjustments to realize their economic potential.

Most people tend to think of a city's economic base in terms of industries. However, the vast majority of people in cities do not have factory jobs; they work in small trades or small informal businesses. We need new measures aimed at making such work viable in the mainstream economy. Young people in the urban underclass are suffering 50 to 60% unemployment. Such high unemployment is potentially explosive, and new policies and approaches are needed; for example, micro-credit for self-employment and for micro-enterprises seems a promising approach.

Extremely destructive social, economic, and political pressures are increasing. A most serious problem is that the rich – whether rich countries or rich individuals – are disengaging themselves, turning their backs on the poor.

...

But perhaps the biggest problem of all is that inequities are growing both within and between countries; these strike at the root of the implicit social contract that enables societies to function, making family and community life possible. According to the United Nations Development Program's *1992 Human Development Report,* if you divide the world's population into fifths, you see clearly that the richest 20% control 83% of the world's income. Furthermore, the richest 20%, which was thirty times as wealthy as the poorest 20% in 1960, is sixty times as wealthy these days.

So the balance of humanity – over four billion people or 80% of the world's population – is living on 17% of the world's income. About 1.2 million are surviving on less than one dollar per day. One billion people have no access to clean water; 1.7 billion live without basic sanitary facilities. The combination of both these deficiencies results in two to three million infant deaths every year due to dysentery or other preventable diseases.

The health impacts of urban environmental problems are staggering. An estimated 1.3 billion people, mainly living in megacities in the developing world such as Cairo, Lagos, and Mexico City, are breathing air that the World Health Organization deems unfit for humans. Less well known is the fact that about 700 million people, mainly women and children, are suffering from indoor air pollution caused by biomass-burning stoves. Such pollution is equal to smoking about three packs of cigarettes daily. In Mexico City, high particulate levels in the air contribute to an estimated 12,500 deaths each year.

Today hundreds of millions of farmers are no longer able to maintain the fertility of their soil sufficiently to eke out a living. Yet during the next few decades the world's food production will need to double to satisfy increased population demand.

Everywhere inequities are also growing, not only within and between nations, but between sectors and occupations. Lawyers and computer programmers, for instance, earn more than carpenters, bricklayers, and janitors, and the differences in income between occupations are increasing. But inequities are also growing within 'high knowledge' occupations. For example, among computer programmers, the gap between the highest and lowest salaries is much bigger than that between the best and worst paid welder or carpenter. Such inequities may owe [something] to the transition from an industrial- to a knowledge-based society. Still, it is unreasonable to think that they can continue to grow without undermining the cohesion of society.

Indeed, almost everywhere there is politically explosive high unemployment that is structural rather than cyclical. In the United States such unemployment is much lower than in other countries. In much of Europe unemployment has remained stubbornly above 10%. In most developing countries, especially in cities and among young people, unemployment often ranges between 20 and 50%.

Imagination is required to devise solutions to these problems. But solutions are feasible. Examples of success and good practice exist. The question is whether we will have the political will to act to make the coming urban century one of promise and well-being, in which the majority of humanity will have the chance to live a decent life.

Source: Serageldin, 1997, p.25

We have talked in this chapter about cities as open intensities. Third World cities are open: apart from anything else they are open to migration from their own countrysides. They are growing fast. Like all cities, then, they are cities of juxtaposition. Cities where different stories meet up.

We also talked earlier about the potential which exists in such places of juxtaposition: of the fact that as a result of their comings-together other things may happen (new cultural forms, as in Mexico, or new economic fortunes, as in Chicago); of the need to turn simple intersection into positive interaction to produce the creative intensity of cities.

What Serageldin's article makes clear is the tension within this potential. The very size of Third World megacities establishes them as major foci of social relations within the world's geography. But these are focal points with the potential either for positive growth or for the generation of despair. Or possibly for both. What can be made of the potential of great agglomerations of people depends on the resources which are available and the conditions within which the city exists.

From Serageldin we can point to some of the conditions within which Third World cities are now struggling:

- growing inequality
- lack of resources
- the pressure to compete
- the fact that the rich of the world are looking the other way.

The question, then, is what kind of urban intensity can be constructed in such a context. Will it be one of (to mention just a few of Serageldin's examples) unemployment, unclean water, a lack of basic sanitary facilities, high infant mortality … ? Or can the cities of the twenty-first century promise 'the majority of humanity' the chance of 'a decent life'?

References

Abu-Lughod, J.L. (1989) *Before European Hegemony: The World System A.D. 1250–1350*, Oxford, Oxford University Press.

Allen, J. and Hamnett, C. (eds) (1995) *A Shrinking World? Global Unevenness and Inequality*, Oxford, Oxford University Press/The Open University.

Allen, J., Massey, D. and Pryke, M. (eds) (1999) *Unsettling Cities*, London, Routledge/The Open University (Book 2 in this series).

Castells, M. (1996) *The Information Age: Economy, Society and Culture*, Volume I: *The Rise of the Network Society*, Oxford, Blackwell.

Jacobs, J.M. (1996) *Edge of Empire: Postcolonialism and the City*, London, Routledge.

Lang, T. (1997) 'Getting food right', *Soundings* Special Issue: *The Next Ten Years,* pp.77–87, London, Lawrence and Wishart.

Lang, T. and Hines, C. (1993) *The New Protectionism*, London, Earthscan.

Lefebvre, H. (1991) *The Production of Space*, (trans. D. Nicholson-Smith) Oxford, Blackwell (first published in French in 1974).

May, J. (1996) 'Globalisation and the politics of place: place and identity in an inner London neighbourhood', *Transactions of the Institute of British Geographers*, vol.21, pp.194–215.

Mehrotra, R. (1997) 'One space, two worlds', *Harvard Design Magazine*, Winter/Spring, pp.25–7.

Sassen, S. (1991) *The Global City: New York, London, Tokyo,* Princeton, NJ, Princeton University Press.

Sennett, R. (1994) *Flesh and Stone: The Body and The City in Western Civilization*, London, Faber and Faber.

Serageldin, M.I. (1997) 'A decent life', *Harvard Design Magazine*, Winter/Spring, pp.40–1.

Townsend, R.F. (1993) *The Aztecs,* London, Thames and Hudson.

Vaillant, G.C. (1944/1950) *The Aztecs of Mexico*, (1944: Doubleday and Doran; 1950: Pelican Books).

Vergara, C.J. (1997) 'One face of the future: Skid Row, Los Angeles', *Harvard Design Magazine*, Winter/Spring, pp.10–13.

Wolf, E. (1982) *Europe and the People without History*, London, University of California Press.

READING 3A
Jane M. Jacobs:
'Negotiating the heart: place and identity in the post-imperial city'

Logic has its limits and ... the City lies outside of them.

(Royal Commission on Local Government in Greater London, 1962)

In the early 1990s the British Law Lords reached a decision which allowed for the development of a relatively small parcel of land located on Bank Junction in the centre of the City of London. Without doubt this is prime real estate. Five major roads intersect at Bank Junction which, among other things, is the site of Mansion House (home of the Lord Mayor of London), the Bank of England and the Royal Exchange. The intersection has the look and the feel of a hub, and its grand buildings stand as monuments to the City's historic centrality to financial and commercial practices in Britain. In 1904 Neils M. Lund depicted the thriving bustle of this city space in his painting entitled *The Heart of the Empire* (Figure 1). Then, as now, the intersection was a symbolic site of a Britain made great by its global reach. Today it is an imperial space in a postimperial age. The decision by the Law Lords effectively ended a development struggle which had begun in the 1960s, which had seen some fifteen years spent in property acquisition, two schemes commissioned from leading architects (neither of whom had lived to witness the fortunes of their designs), and two gladiatorial public inquiries in which heritage redevelopment as opposed to new build development was the central issue.

The struggle over this symbolic heart of empire resulted in a prime piece of real estate being locked out of one of the most rapid and dramatic periods of restructuring and property speculation ever seen in the City and its surrounds ...

In this prolonged redevelopment saga, Bank Junction (past, present and proposed) was invested with meaning by a range of interest groups: the developer, the local authority for the area (the Corporation of London), various local businesses and, not least, the powerful conservation lobby groups. The discourses generated by the planning controversy were not simply about the form and the function of this section of the City of London, a battle of old versus new or small business against big business. This highly publicised planning controversy became a nodal point in the imaginative reaffirmation of the identity and status of the City in relation to the nation and the rest of the world. In the span of half a century the City of London had gone from being the centre of an empire with a global reach, to one of the few urban centres given the privileged designation of 'global city' (King, 1990, 1991; Sassen, 1991). The City of London of the 1980s was both a postimperial city and a 'postmodern(ising)' city. The City had moved from the confidence afforded by empire to a more competitive and at times precarious status constituted out of new global and regional alignments.

... Robins (1991, p.23) argues that 'Empire has long been at the heart of British culture and imagination'. Certainly in the City of London the idea of empire is not confined to the past. It is an active memory which inhabits the present in a variety of practices and traditions and which still works to constitute the future of the City. Place plays an important role in the way in which memories of empire remain active. For example, the efforts to preserve the historic built environment in the present are often also efforts to preserve buildings and city scenes which memorialise the might of empire. Imperial nostalgias are not tied just to preservationist interests. New build schemes, such as the ones proposed for Bank Junction, may also activate a

FIGURE 1 *Neils M. Lund's 1904 painting,* The Heart of the Empire, *depicts Bank Junction as the monumental, thronging hub of nineteenth-century imperial might. The painting takes an aerial perspective from above Hawksmoor's St Mary Woolnoth and looks westward past Mansion House (on the left) towards St Paul's Cathedral. The triangular group of nineteenth-century commercial buildings visually link St Paul's to Bank Junction by appearing as an extension of the Cathedral itself. These are the very buildings that were demolished during redevelopment. (Reproduced by courtesy of the Guildhall Art Gallery, City of London)*

memory of empire even while they display a seeming disregard for its built environment legacy. Furthermore, both heritage and new build schemes are active place-making events in which social and economic visions are articulated. Imperial nostalgias, then, work through place in a multiple register. They are present in schemes to preserve what was and also in visions of what might be. But they are also present in the often discordant resonance between such place-making events and economic and social orders, real and imagined. The heritage battle of Bank Junction is about how an activated past assists in the City's (and the nation's) adjustment to the loss of empire.

... [I]n the 'new world order' Britain has forged new global and regional alignments. The clearest of these, and the one most pertinent to the politics of identity and place which operates in the contemporary City of London, is what might be thought of as Britain's postimperial return to Europe. In the contemporary City of London, imperial nostalgias cohabit with the imperative of a creating regional alliance with Europe.

Difference gathered in the City of London

... In the City of London of the 1980s ... two distinct narratives of place ... had begun to gain popular currency in relation to the City of London. On the one hand, the City had been singled out by commentators like Saskia Sassen (1991) as a paradigmatic 'global city' ... In so describing London, and in particular the finance-centred City, such an explanation rests on the internationalisation and expansion of the finance sector and the trading in services, as well as a re-patterning of foreign investment facilitated by deregulation and transformed communications technology. The City of London is a space given over to finance and business. London, it seems, had successfully shifted from the global geography of empire to the global geography of 'transterritorial markets' (ibid., p.327).

On the other hand, the City of London of the 1980s became implicated in a new urban design movement which advocated a commitment to indigenous architectural forms and a domestic, village-like townscape. Prior to the ... more sensational and personal figuring of the Prince of Wales in the media, his public profile was linked to more sedate interventions in planning and architecture. His views on architecture and planning were brought to public attention through a Victoria and Albert exhibition, a television documentary and a book, all entitled *A Vision of Britain* (HRH The Prince of Wales, 1989). The Prince's architectural programme is framed as deferentially indigenous, bowing to the natural and organic character of British architecture and townscape. His 'ten commandments' of architectural design are presented modestly and mystically as if pieces of folklore – 'non-expert' views excavated from an architectural wisdom which springs forth from the land (ibid., pp.75–153). Wright (1991, p.239) suggests that Prince Charles 'makes his nation sound like a land of white bushmen'. His visions are at once both local and national in their reverberation; celebrating the local in order to restore a certain architectural order to the Kingdom. The defence of the local, the community, acknowledges the

rights of ordinary places and people but in the service of a uniquely British (some would say English) scene and nation. The principles of the Prince of Wales contain a certain anxiety about a loss of order and are as much about nostalgia for stability as they are about the less manageable possibilities of local empowerment. He draws upon a wide range of architectural examples to furnish his polemic, but the City of London, and particularly the postwar transformation of its Canaletto skyline of church spires, is central to his argument about the demise of vernacular urban design more generally (see Daniels, 1993, pp.12–13).

Making monuments

The thirty-year planning controversy surrounding the redevelopment of this site on Bank Junction has the distinction of being the longest running planning battle in London's history. Since the early 1960s the developer Peter Palumbo, through his property development firm, City Acre Property Trust, had been acquiring a group of Victorian buildings on Bank Junction with a view to placing on the site a prestigious office development of the highest architectural quality. He presented two main schemes. The first scheme, commissioned in 1962 but not presented for formal planning approval until 1982, was an eighteen-storey modernist office tower designed by Ludwig Mies van der Rohe and grandly titled the Mansion House Square scheme (Figure 2). The second scheme was a five-storey postmodern office development designed by James Stirling of James Stirling, Michael Wilford and Associates and called the No. 1 Poultry scheme (Figure 3). Both proposals required the demolition of a block of Victorian buildings, eight of which were listed as historic buildings, and all standing within a designated Conservation Area (Figures 4, 5 and 6). Operating through the protection provided for under listed building and Conservation Area legislation, the local authority for the City, the Corporation of

FIGURE 2   *The modernist Mies van der Rohe scheme for Bank Junction was initially commissioned in the 1960s and addressed the surrounding townscape in name only*

London, decided to refuse planning permission for both schemes on the grounds that they would seriously damage the historic character of the area. Using his statutory rights under existing planning legislation, Palumbo was able to challenge the local authority decision by asking the Secretary of State for the Environment to 'call in' the decisions for contestation in the arena of the public planning inquiry.

The proposals to redevelop part of Bank Junction went to public inquiry twice during the 1980s: the Mies tower in 1984 and the Stirling scheme in 1988. In the first inquiry the Mies tower was refused planning permission, but Palumbo was encouraged to submit a new scheme. The No. 1 Poultry scheme was commissioned but again went to a public inquiry. The findings of the second inquiry were against the granting of planning permission, but this recommendation was over-ruled by the Secretary of State for the Environment and planning permission was granted. The decision was challenged by conservationists and it was only through appeal to the Law Lords that permission to redevelop Bank Junction was finally secured.

The Corporation of London's refusals to grant planning permission were in part a response to obligations and options provided for under conservation legislation. Yet the Corporation's responses to Palumbo's visions were far from consistent. In the late 1960s, when Palumbo first mooted his vision for a modernist office tower on this central site, he was granted provisional planning permission. It was only when Palumbo applied for full planning permission some fifteen years later, after he had finally secured ownership of the majority of the development site, that he faced Corporation resistance in the form of the refusals to grant permission to build.

FIGURE 3   *The James Stirling and Michael Wilford scheme for Bank Junction – No. 1 Poultry – adopted a scale and style inspired by surrounding buildings*

FIGURE 4   *The group of mainly nineteenth-century buildings on the Bank Junction development site*

FIGURE 5   *The void left after demolition*

141

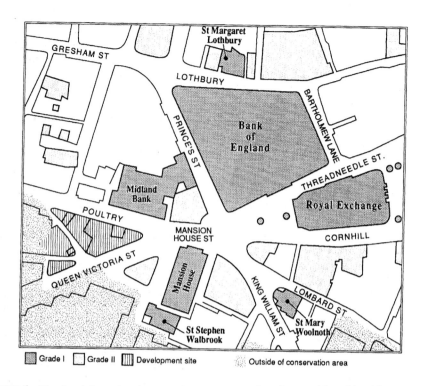

FIGURE 6 *The Bank Junction intersection is encrusted with grand listed buildings:
Soane's Bank of England, Dance's Mansion House, Tite's Royal Exchange and Lutyens'
Midland Bank are the four most imposing buildings on the intersection. Slightly set
back from the intersection is Hawksmoor's St Mary Woolnoth and Wren's St Stephen
Walbrook ... The area is entirely overlain with Conservation Area status which
recognises the historical value of the individual buildings but also the townscape value
of the buildings in relation to one another*

Picturing the Empire

When Palumbo applied for full planning
permission to develop his Bank Junction
site in the mid-1980s, conservationists and
the Corporation of London rallied to the
defence of what heritage authorities had
designated as 'a national architectural set
piece' (English Heritage, 1988). Those
opposing demolition and new build
development were, at one level,
concerned with the loss of individual
buildings that had historical merit and
listed building status. But there was an
equally strong concern for the collective
value of the buildings on the development
site and their relationship to the
surrounding area. This was a concern for
the townscape quality of the area and, in

particular, the visual relationship between
the more diminutive Victorian buildings
on the development site and the
surrounding cityscape. The Corporation of
London engaged the services of
townscape expert Roy Worskett to prepare
and present its case for refusing planning
permission to the scheme.

Since the 1970s the Corporation of
London has vigorously pursued the
conservation of the historic townscape
quality of the City ... The 1986 Local Plan
for the City was emphatic in its
endorsement of the conservation/
townscape approach to planning. The
'architecture, skyline and distinctive
townscape' of the City were all to be
'preserved and enhanced' (Corporation of

London, 1986, p.126). This has been implemented through restrictions on building heights, style guidelines, and the encouragement of refurbishment as opposed to demolition and new build.

The incorporation of the idea of 'townscape' into local City of London planning accorded with more general trends which shift the emphasis of conservation away from individual buildings. Townscape is concerned with the visual perception of the urban environment in compositional pictorial terms: viewing cities as similar to paintings, 'as problems of composition, based on the production of a series of harmonies or contrasts ... the city as visual art' (Anderson, 1988, p.405). The key emphasis in such assessments is 'serial vision': the way in which elements of the urban scene interact visually as the observer moves through space. Townscape policy cherishes 'informality'. 'accident' and 'spontaneity' but its creation and maintenance are contrived through active intervention in the urban scene, either through conservation or through the addition of certain built forms ... Townscape is now a commonsense notion in British and other planning systems. Ironically, the translation of the townscape idea into urban policy has created a planning regime which regulates for acceptable visual 'disorder'.

Townscape as an approach to planning was initially developed and promoted by the editor of the British periodical Architectural Review, Hubert de Cronin Hastings. He campaigned for a 'visual policy' of urban landscape, drawing on the eighteenth-century rural picturesque, which, in his view, was 'that landscaping tradition to which England owes its most personal aesthetic character' (de Cronin Hastings, 1944, p.5). The townscape concept was later given broader planning popularity through the writings of Gordon Cullen, one of the regular writers for the Review, who published a formal set of townscape principles (Cullen, 1961). For Hastings, the English city was

characterised by its 'infinite variety' and it was the task of planning to embolden 'irregularly' and 'disdain formality'. Hastings saw the responsibility of the planner to be the enhancement of inherited, 'natural', visual disorder – a state he dubbed ' "sharawaggi", after an "East Asian" term for irregular gardening' (de Cronin Hastings, 1944, p.5). This was an argument for the improvement of a 'scene according to the manner suggested by itself', a notion of development based on the *genius loci* of place, the intrinsic, indigenous qualities of the local.

The development and promotion of the townscape idea was set in direct contrast to modernist ideas of planning and architecture which had emerged on the continent in the inter-war years and had begun to appear in Britain. Townscape, Hastings argued, was more compatible with the English spirit and aesthetic and the English appreciation of 'age and quaintness' (de Cronin Hastings, 1945, p.165). Writing as 'I. de Wolfe' in 1949, Hastings proposed that this urban design policy would ensure that there was a uniquely English 'regional development of the International Style' (de Wolfe, 1949, p.355) ...

Matless (1990) has argued for the case of rural England, that processes of ordering and orchestrating 'indigenous disorder' are as deeply connected to the presentation of Englishness as the straightforward process of preserving what is old. In the controversy over the redevelopment of Bank Junction, the townscape assessments of the development site and its surrounds elaborated the status of this place as the symbolic heart of empire. And just as the idea of townscape emerged and has lived on through a tension with notions of 'foreign' continental European sensibilities of planning, so too did the assessments of townscape in this 1980s planning struggle express a domesticated memory of empire constructed in opposition to a demonised European other.

Pleasures of hearth

On the walk westwards along Cornhill into Bank Junction there is a short section of about fifteen paces where the dome of St Paul's looms in the skyline. In the Corporation's argument against the No. 1 Poultry redevelopment, this glimpsed view of the dome from Cornhill was claimed to be the 'most striking and significant aspect' of the Bank Junction area (Worskett, 1988, p.4). The No. 1 Poultry development proposal all but obliterates this glimpsed view, leaving only the cupola visible (Figure 7). In the Corporation's case, this existing view of the dome was considered nothing without the supplementary visual effect of the buildings of Bank Junction and particularly those on the appeal site. The Mappin and Webb turret of the existing Victorian buildings 'framed' and 'played' with the dome producing a 'superb kinetic view' (ibid., p.54).

The conservationist's defence of the glimpsed view of St Paul's dome from Cornhill is an extension of a long-held reverence for the visual supremacy of this great architectural piece of the City. St Paul's was the edifice of Wren's rebuilding of the City's churches after the Great Fire of 1666. Its status as a symbol of City survival gained new potency on the night of 29/30 December 1940, when the City faced one of its first direct German attacks of the Second World War. Almost one third of the City's fabric was destroyed in that and subsequent bombing raids. While the area immediately north, east and south-east was devastated by wartime bombing, the dome of St Paul's remained virtually intact. The image of the dome under attack from the Germans became a symbol of heroic British survival (Daniels, 1993). Throughout the Corporation's postwar plans to remodel and rebuild the City there was continual reference to the need to protect and enhance the visual dominance of St Paul's. As early as 1934 the Corporation undertook its first study on height control in relation to St Paul's

dome and restrictions became policy by 1935 (Kutcher, 1976, p.161).

Today, views of the dome of St Paul's, both from afar and from within the City, are marked and protected ... The glimpse that might be had of the dome from Cornhill is the only view from the 'heart of the City' itself. For opponents of the No. 1 Poultry scheme this marked a special link between the eighteenth-century dome, the nineteenth-century city of empire and the present:

> [the view from Cornhill] is not just a view of St. Paul's from afar. It is the relationship between Bank Junction, Mansion House and the Mappin and Webb triangle and the metropolis and Empire ... [T]his viewpoint is ideal to give a sense of London as the economic centre of the Empire as well as the spiritual and other-worldly sense of the Empire.
>
> (English Heritage, No. 1 Poultry cross-examination, 1988)

Here the economic worldliness of empire, as embodied in the monuments of Bank Junction, is offered present-day moral absolution by the 'other-worldly' influences of St Paul's. Joseph Conrad wrote in *Heart of Darkness* (1902) that imperialism is 'not a pretty thing' and that its only redeeming feature is the 'idea at the back of it'. The battle of Bank Junction was also a struggle to preserve a spiritually redeemed heart of empire. In this struggle, the 'idea of empire' was once again taken away from its grimy past and placed into the foreground of a pretty scene in the present.

In the townscape argument against the No. 1 Poultry redevelopment, concern was also shown for the more immediate visual relationship between the existing group of Victorian buildings on the development site and other buildings abutting the Junction, such as the Bank of England, Mansion House and the Royal Exchange. The diversity and smallness of scale of the existing buildings on the development site were seen as relating positively to the

FIGURE 7 *The local view of St Paul's from the narrow street of Cornhill would be blocked by the proposed development. Opponents argued that this was a breach of existing local regulations to protect views of the dome. The developer argued that the No. 1 Poultry scheme would open out new views of the dome while the circular drum at the centre of the building would echo 'in absence' the proportions of the Cathedral dome.*

monumentality of the surrounding buildings. The expert witness for the Corporation of London likened the relationship between the more humble Victorian buildings and the grander surrounding buildings to a 'theatrical show' in which the Victorian buildings were the 'supporting cast' to the 'stars' (Worskett, 1988, p.56). The 'visually subservient' nature of the buildings on the development site was seen to be their most important contribution to the character of the area and was likened to 'the relationship of visual master and servant' (ibid., p.39).

This recourse to the notion of urban hierarchy is a common feature of townscape rhetoric (Cullen, 1961). It is presented as a benign ordering which provides visual diversity, a type of grammar necessary for the correct comprehension of urban form. However, the townscape concept is, like any other form of landscape idea, a social

construction which naturalises the operations of power (see Daniels, 1989). The hierarchical 'good manners' embodied in the principles of townscape reach beyond their immediate point of reference in the built environment of Bank Junction and address desired social orders. This broader resonance between the local townscape argument and an idealised social order is evident in what were then very public and quite influential views of the Prince of Wales on urban design. The Prince of Wales considers architectural hierarchy as especially important and includes it as one of his ten principles of planning. In advocating the need for urban hierarchy, the Prince makes a direct reference to the social order for which it is supposed to stand. In his view townscape hierarchy is common sense since 'civilized life is made more pleasurable by a shared understanding of simple rules of conduct' (HRH The Prince of Wales, 1989, p.80). Here one senses a nostalgia that extends beyond the heritage value of the built form, to a social and moral order once more surely held by the nation and reminiscently embodied in this symbolic site of empire.

Those contesting the proposed redevelopment turned to a particular image as incontrovertible evidence of the wisdom of their view. Hanging in majestic and nostalgic proof of evidence at the rear of the Public Inquiry room was Neils M. Lund's turn-of-the-century painting entitled *The Heart of the Empire* (see Figure 1). In this painting, Bank Junction is the visual hub of a city which seethes with the life of empire. Here is a place shot through with a vibrancy that had derived, quite literally, from the wealth obtained from exchanging raw materials from the colonies for manufactured goods in the core. The diminutive Victorian buildings of the development site play a crucial visual role in Lund's painting. They act as the visual pivot of the scene, an extension of the dome of St Paul's guiding one's gaze to the dome and then beyond to the edge of empire. The battle of Bank Junction activated the memory of the City at a time when ... [it] felt secure as the centre of a global empire and where the social relations, both at home and abroad, were more surely set in its favour. Through the townscape argument the local planning authority and conservationists were constructing a symbolic terrain which spoke of values of morality, civility, hierarchy and order once central to the City of empire.

Imperial illusions

During the postwar years the City has shifted from an empire internationalism, with financial business tied to colonial modes of production and trade, to a new global internationalism (King, 1990, pp.9, 83–87). The Euromarket (that is trading in currencies held off-shore from the country of origin) has been critically important in the transformation to a global city (Plender and Wallace, 1985, p.26; King, 1990, p.91). For example, between 1965

and 1981 the size of the City's Euro-currency market had risen from US$11 billion to US$661 billion (Stafford, 1992, p.33). This market ensured that the City maintained and adjusted the source of its financial dominance in international terms. Not all sectors of the City enjoyed such expansion and in particular the operations of the Stock Exchange were seen to be hindering the City's capacity to compete effectively with other financial centres. Change came to the Stock Exchange in October 1986, in the form of the 'Big Bang'. Fixed commissions were abolished, allowing for single capacity trading, that is, brokers/dealers acting both as agents for others and on their own behalf in the buying and selling of stock (DEGW, 1985, p.8). The unlimited liability requirements, which had previously limited the companies that could join the Exchange, were also lifted. This 'deregulation' was accompanied by a major transformation in the technology and communications base of the financial sector. Screen-based trading was introduced which in some sectors has facilitated twenty-four-hour global trading. The impact of deregulation and the new technology on the financial sector was marked. Turnover in equities, for example, increased from an average of £650 million per day before the Big Bang to over £1.1 billion per day post-Big Bang (Clarke, 1989, p.125). The heady times following the Big Bang were tempered by the 1987 October Crash. The FTSE fell a record 250 points and it was estimated that some 3,000 City jobs were lost. Despite the Crash, such had been the impact of the international-isation of banking and trading, deregulation and new technology, the City remained buoyant in its new-found 'global' status.

These various restructurings placed the pressure of change on existing social organisation and business practices within the City (Harris and Thane, 1984; Budd and Whimster, 1992). The City of London has always been simultaneously cosmopolitan and British (Cain and Hopkins, 1993, p.27). It has always accommodated difference but within the limits established by the class, race and gender-specific sociology of its financial practices: the stereotype of the pinstripe-suited, Oxbridge-educated, businessman operating through intricate structures of liveries and clubs (Lisle-Williams, 1984; Cassis, 1988). During the 1980s this surety was being shaken by a rapid phase of change which entailed new investment players, new financial institutions and conglomerates, new practices, and a changing and expanding labour force. For example, an emergent Euro-bond market in the City encouraged the growth of non-British banking interests. In 1914 there were only thirty foreign banks; by the 1930s this had expanded to over eighty and by the early 1960s there were over one hundred (Goodhart and Grant, 1986, p.9). From 1961 to 1971, the number of foreign banks in the City doubled and in the following decade doubled again. In 1987 there were 453 foreign banks either directly or indirectly represented in the City (King, 1990, p.89). Thrift and Williams (1987) and Thrift (1994) have sketched the restructuring of the City labour force associated with the Big Bang and noted the ways in which it expanded rapidly and was associated with high pay and accelerated rates of salary growth. The managerial sector of the City's new workforce may have continued to conform to traditional class formations, but it is likely that the influx of younger workers to the City workforce transgressed familiar boundaries of class, race and gender (see also McDowell and Court, 1994).

The restructuring of the City's traditional financial practices had a specific geographical expression. The new practices of the financial sector were increasingly dependent on computer-based trading and information technology and this generated demand for entirely new building types with different floor to ceiling heights to take cabling as well as large open internal layouts for conversational trading (Pryke, 1991). There was also increased demand for high-quality office space in buildings with an architectural style that could be linked to the corporate identity of its occupants. This expanding and increasingly specialised demand for office space could not be met by existing City stock. In the six-month period following the Big Bang of 1986 the availability of office space fell by some 23 per cent (Richard Saunders and Partners, 1986). Rents reflected the scarcity of suitable office space and in mid-1988 rents in the City centre had reached their all-time high of £60 per square foot, with the occasional rental of £70 per square foot. The way was open for a property development boom directed at meeting the new demand.

Palumbo argued for his development proposals on the basis of the global status of the City and this accelerating demand for quality office space. He may well have expected the local planning authority, the Corporation of London, to respond positively to such logic. The Corporation had not resisted the pressures for change in the City. While in 1982, for example, planning permission was given to just 689,000 square feet of office development, in 1987, the year following the Big Bang, the Corporation issued permits for over 12.5 million square feet of office floorspace either in the traditional core or on perimeter sites under its control (Corporation of London, 1993a, p.3). The mood of the local authority seemed to favour development. But the development ethos of the City had limits that were tied to the Corporation's view of how the City was best able to maintain its status as a 'global' centre.

The Corporation of London sought to differentiate itself from other competitors. The 1986 City of London plan elaborates:

> The City of London … is noted for its business expertise, its wealth of history and its special architectural heritage. The combination of these three aspects gives the City a world-wide reputation which the Corporation is determined to foster and maintain … The City's ambience is much valued and distinguishes it from other international business centres.
>
> (Corporation of London, 1986, p.3)

The preservation and enchancement of the local character of the City was seen as the very 'underpinning' of its global status and as an attraction to growth rather than as a deterrent. Heritage, correctly preserved and enhanced, was seen as the way the City could promote itself as distinctive in a new global market and allow it to compete effectively against challenges arising closer to home in the form of the Docklands redevelopments, the possible rise of Frankfurt as a centre of a unified European financial community, as well as its 'global' competitors of Tokyo and New York.

The new communications technology associated with the 1980s restructuring of the City allowed an unprecedented spatial flexibility (Pryke, 1991). Proximity to the core, which had long determined the spatial patterning of the City, was no longer necessary. Much of the post-Big Bang speculative development was located on or beyond the outer edges of the City, where development speculators could take advantage of large sites which had become available because of various closures and restructurings in the transport and manufacturing sectors ... The second redevelopment scheme for Bank Junction offered a modest 125,000 square feet of lettable office space at a time when some 8 million square feet of City office construction was underway and planning permissions had been granted for a further 7.5 million square feet (Valuation Office, Inland Revenue, 1988, p.29). The Canary Wharf and other proposed Docklands Enterprise Zone developments added to this surfeit almost 12 million square feet of office space (London Planning Advisory Committee, 1993, pp.45–56).

Without doubt, the conservation-mindedness of the local authority and its regulation of change to the built fabric of the City contributed to the diminished opportunities for development within the core areas of the square mile, so accelerating the City's outward push. New build development of the scale required by transformed City business practices was simply not available in the City core,

which was almost entirely covered by designated Conservation Areas and where building up, as at the Canary Wharf development, was not possible within the strict height restrictions of local planning policy. As one spokesperson for the developer stated, a 'new business heart to the City' was being created on the border while 'the traditional heart is frozen as an historic monument' (Baker Harris Saunders, 1988, p.9). While it is easy to read the conservation tendencies of the local authority as a loyalty to the local vernacular, its effect may well have been to destabilise what it sought to preserve. It was not simply the appropriating forces of global capital (or the opportunities of new technology) that were decentring traditional City geographies. It was also the urge to preserve an historic built form in a context where new development opportunities were afforded by the opening up of land on the edge of the City.

By the time Palumbo's vision of the 1960s could begin to materialise in the 1980s and finally gain Law Lord approval in the early 1990s, it was not only his Mies van der Rohe office tower that seemed out of date, but his entire development strategy. This included his costly loyalty to the heart of the City as well as his unusual status as a lone developer in an urban centre under massive speculative development pressure from a new breed of property investment and development consortia (Healey, 1990, p.9; Pryke, 1994). Furthermore, the last stage of Palumbo's struggle to redevelop this prime piece of real estate spanned a time when the property market went from boom to almost bust ... It was in this [later] climate of contraction that Palumbo's development proposal finally gained planning approval from the Parliamentary Law Lords. Not only had Palumbo been locked out of the boom period, he now had to actualise his dream in a contracting development phase.

The developer of Bank Junction and his opponents may have had different visions for the site but they shared a loyalty to this place as the symbolic heart of the City. This can be seen in the developer's shift from the uncompromisingly modernist Mies van der Rohe scheme to the post-modern style of James Stirling's No. 1 Poultry scheme. This was a strategic effort by Palumbo to adjust his development impulse to the architectural and planning sensibilities of the local setting in order to realise his development ambitions in the historicised inner core. The new No. 1 Poultry scheme conformed to local height restrictions. The development was largely confined to the existing street pattern. The design itself echoed the existing built form. Those advocating the No. 1 Poultry scheme argued that the design was inspired by local context and the historic character of the core. Stirling's design, his expert witnesses argued, was 'equal' to and 'did justice' to the surrounding monumental buildings of Bank Junction. While the scheme may have blocked the views of St Paul's from Cornhill, the central drum of the building echoed the dimensions of the dome and the tower had a viewing platform that would offer views of it never before seen. This design self-consciously sought to mimic all the qualities of the Junction which gave it its special place in the imperial nostalgias of the present ...

Continental entanglements

The constitution of identity is not only marked by an inward turning to place. Of equal importance is the marking of difference through the notion of a feared Other. This marking of difference became the means by which the varying interests involved in the struggle to redevelop Bank Junction, actually unanimous in their loyalty to the importance of this place and the City, engaged in the performance of opposition. For Palumbo's opponents, the development schemes themselves were 'alien' to the local scene simply by virtue of being new buildings. But the full weight of his 'alien' status was elaborated by ascribing to the schemes broader social categories of Otherness.

... Britain now figures only faintly as part of the Europe of ... colonisation. Its position in relation to the centre of the reconstituted European ... economic union is also uncertain. In 1992, the year in which Europe struggled towards the finalisation of the Maastricht Treaty, a Eurosceptic writing in the Right-wing British magazine *The Spectator* proudly described Britain as 'a stroppy trading nation on the margins of Europe, ever striving to keep clear of continental entanglements' (Davies, 1992, p.12). Britain may well have been the pulsing hub of an empire, but the nation had an ambiguous and at times antagonistic relationship with continental Europe – particularly its Germanic core. The moves towards European economic and political union have meant that once again Britain faces its continental Other. This encounter between Britain and the Germanic centre of Europe manifested itself in the struggle to redevelop Bank Junction. Here the 'alien' nature of the proposed developments was elaborated through reinventing them as 'German' and thereby activating a familiar narrative of German antagonism towards Britain.

The first Palumbo proposal for Bank Junction, the modernist Mansion House Square scheme (see Figure 2), had an explicit German connection. The architect, Mies van der Rohe, was German-born. But the foreignness of Mies and his Mansion House Square scheme was not tied in a simple way to the architect's place of birth. In Mies's early years as an architect in Germany he fell out of favour with Hitler and fled to the United States. Indeed, it was not until the 1960s that Mies returned to Germany with a commission to design the National Gallery in Berlin. But in true International Style, the design used for this national monument was one originally commissioned for the unbuilt headquarters of Bacardi Rum in Santiago, Cuba (Balfour, 1990, p.224). Mies's design for Bank Junction was criticised by opponents not because of the architect's tenuous Germanic links but more because

Mies, and his architecture, claimed to 'transcend the barriers of nationality' (Marks, 1984, p.39). As one of the advocates of the Mansion House Square scheme stressed, Mies was 'not an American architect, nor truly a German one, but an international architect in every sense' (Pawley, cited in ibid., pp.49–50).

Mies's architecture is recognised as one of the purest forms of the 'International Style' to emerge from inter-war, continental Europe … In his architectural philosophy it was possible to project an 'ideal space' on any place (Balfour, 1990, p.55). For Mies, the only 'real' architecture was that which 'touches the essence of the time' – which, for him, was not history but a transcendent order of geometrical reason (Mies van der Rohe, cited in Johnson, 1978, pp.203–4). Opponents of the Mansion House Square scheme noted the 'curious' lack of any 'sense of place' in Mies's work. And the Corporation of London pointed to Mies's lack of familiarity with the City: he had visited the City only twice before presenting his scheme (Corporation of London, quoted in Marks, 1984, pp.50–51).

It was this lack of sensitivity to the local that enabled Mies's tenuous Germanness to be reasserted and ascribed to the Mansion House Square scheme. The Prince of Wales's architectural vision provided the lead for this translation of a diminutive Little Englandism to a combative, xenophobic fear of the German Other. The Prince of Wales was outspoken about the Palumbo proposals for Bank Junction, labelling the Mies tower a 'glass stump' (HRH The Prince of Wales, 1989, p.66). More specifically, he likened the destruction that Palumbo's visions would wreak upon the City townscape as akin to that of the Luftwaffe (HRH The Prince of Wales, 1987, quoted in Jencks, 1988, p.47). Conservationists and architectural new traditionalists in Britain consistently earmark the Blitz as the beginning of the end for British architecture (Amery and Cruickshank, 1975; Esher, 1983). The wartime destruction of the City opened the way for

massive reconstruction, much of which was executed in the International Style, with a clear lineage to the architects who emerged from 1930s continental Europe. This 'alien' architectural idiom is construed as much as an invasion of British soil as the Second World War air raids (HRH The Prince of Wales, 1989, p.9). In a publication advocating townscape planning, Tugnutt and Robertson (1987, p.16) argue that modernist planning doctrines, 'largely imported from Europe', effected what they call a 'second Blitz'.

By 1988, when the No. 1 Poultry scheme was facing public planning inquiry, a further round in the political and economic restructuring of Europe was underway. The discursive constitution of Palumbo's second development scheme was set within a period marked by heightened British concern about European political and economic union and, in particular, fears of German ascendancy in the new Europe. The move towards Monetary Union has accentuated British concerns about German ascendancy in the new Europe. These are fears hard felt in the City where the specific anxiety is that it may relinquish its current dominance of the financial sector to Frankfurt, home of the Bundesbank, and a respectable financial centre in itself. A recent Corporation of London survey suggests that perceptions of European, specifically German ascendancy, are somewhat misplaced. In most of the major financial activities, the City remains in a strong position. Yet the survey report notes that many City interests perceive Frankfurt to be 'by far the biggest threat' because it is actively pursuing some of the financial business traditionally centred in the City (Corporation of London, 1994, p.27). The Corporation report even makes the tellingly faint comparison that 'more people work in financial and business services in Greater London (over 700,000) than the entire population of Frankfurt (approximately 600,000)' (ibid., p.19). According to this report a worst-case scenario for the City would involve a single European currency which excluded sterling and which would be run by a German-based central bank (ibid., 29).

...

It might be expected that the self-consciously 'local' design logic of the British (Scottish) architect James Stirling (see Figure 3) would evade charges of 'foreignness'. Yet Stirling's No. 1 Poultry scheme was not only cast as foreign but, in the context of the growing British 'crisis' with a restructuring Europe, its alien status was again ascribed by way of a specified 'Germanness' which drew upon the evocative war metaphor. The No. 1 Poultry scheme was described by opponents as 'aggressive', even 'militaristic' (Worskett, 1988, pp.160–1). In the planning inquiry James Stirling was subjected to cross-examination more like a wartime treason trial:

> *Corporation of London:* You say that the prow does not overpower Mansion House, but is it not reminiscent of a German defence works?
>
> *James Stirling:* No. I notice you refer not to English bunkers but to German ones.
>
> *Corporation of London:* I am not saying German in a derogatory way ... German bunkers are more powerful.
>
> *James Stirling:* You obviously know German bunkers!
>
> (Personal transcript, No. 1 Poultry Inquiry, 1988)

But it was actually James Stirling who was proven to have the more initimate knowledge of this building type. Under pressure of cross examination, Stirling 'confessed' that he had in fact been involved in modifying such defence bunkers (albeit British ones) after the war. Even the supposedly impartial public inquiry Inspector found it difficult to resist the appeal of this line of questioning. He brought to the inquiry a book on the buildings of the Channel Isle of Alderney, and showed a photograph from it of an 'Alderney Eyesore'. It was a German control tower, and legal advocates on both sides conceded that there was a striking similarity in style between this structure and the Stirling proposal (ibid.). Opponents of the development argued that while Stirling was acclaimed as one of Britain's 'big three' architects one of his most renowned buildings is the Neue Staatsgalerie in Stuttgart ... and that some of his better known schemes in Britain were infamous for their design faults. Palumbo's opponents cast James Stirling as a 'traitor' who had an architectural style better suited to the taste and disposition of a demonised German Other. Palumbo's vision to redevelop was cast as a act of national subversion – an attack by the German core of Europe on the very heart of the City.

...

... The lengthy and costly planning struggle Palumbo faced in seeing his vision materialise created an opening for the realisation of the fears of his opponents. Thirty years of development dreams, architectural schemes and legal costs meant that when final planning permission was received from the Law Lords, Palumbo needed to seek co-investors. Late in 1993 Palumbo secured a 50:50 joint venture partnership with Dieter Bock, chairman of the Frankfurt-based company Advanta AG. Bock's private partnership with Palumbo was to ensure that a scheme now estimated to have a development cost of £50 million would finally come to fruition. The construction of No. 1 Poultry is to be jointly financed by Advanta and Palumbo and, on completion, is to be jointly let and owned by the two parties (*Architects' Journal*, 22 September 1993, p.10). In one report readers were reminded that while the Prince of Wales likened the Stirling design to a giant '1930s wireless', the same scheme was not only financed, but also 'admired', by Dieter Bock (*The Independent*, 22 September 1993). This final stage in the development saga may well speak to a more general trend in City property investment. A recent chartered surveyor's report based on all property deals in the City of London notes that German investment in City property is around 38 per cent compared with a UK investment of 39 per cent, the remainder given over to other foreign investors (Jones Lang and Wooton, 1994, p.4). Perhaps the feared 'invasion' is already underway and the beloved local of the City of London is 'inhabited', in a most material sense, by the Others of the global.

Colonial returns

In April 1992 the IRA detonated a bomb in St Mary Axe near the Baltic Exchange, killing three people and causing considerable damage to nearby buildings, particularly the Commercial Union Building. A year later it bombed a site near the junction of Wormwood and Camomile Streets, badly damaging the London headquarters of the Hong Kong and Shanghai Banking Corporation and blasting windows from one of the City's tallest buildings, the NatWest Tower. The second bomb produced an overall damage bill of £1 billion, as well as killing one person and injuring forty-two others.

Northern Ireland is the oldest, most tenaciously monitored and, in these formally postcolonial times, decidedly idiosyncratic remnant of British imperialism. During the period in which formal British involvement moderated and, eventually, evaporated in most of the empire, intervention in Northern Ireland intensified. Attacks like those on the City of London were not a new feature of IRA activity and over the past few decades the IRA has frequently bombed sites in England including, of course, the notorious bombing of the Tory Party Conference in Brighton in 1984.

In choosing the City of London as its target in this new round of military activity the IRA was not simply attacking the symbolic heart of empire, it was also attacking the precariously placed Britain of the New Europe. The second bomb was timed to coincide with the annual meeting of the European Bank for Reconstruction and Development which attracted over 1,000 businessmen and bankers from around the world. One City banker

remarked that the IRA 'could not have picked a better day to damage the City's reputation' (*Sunday Times*, 25 April 1993). The media coverage of the bombings made much of the possible harm this spectacular display of the City's vulnerability would have on its chances of being chosen as the site of the European central bank. City interests were more stoic in their response and assured the world it was 'business as usual ... for the financial and commercial heart of Britain' (Corporation of London, 1993c).

The Corporation of London's immediate response to the second bombing was tactlessly consistent with its over-riding conservationist agenda. The Planning Officer for the Corporation suggested that the damaging of the NatWest Tower provided the opportunity to demolish the building and replace it with something more suited to the City's character. As it is likely that both bombs were carried into the City in large vehicles, he also suggested that the bombings revealed the need for more rigid traffic restrictions in the City (*Sunday Telegraph*, 2 May 1993). The bombings, which the Corporation acknowledged were a direct result of the City's role in financial activities, suggested to the Corporation that it accelerate its thinking on 'making the City less vulnerable as an economic target' and on making it a more 'livable' place which included residential space, pedestrian precincts and restricted traffic flows.

Although the IRA attacks on the City were conceived as acts of anti-colonialism, it was a most un-English City that suffered the direct force of the bombings. The full force of the second blast was most severely felt by the Hong Kong and Shanghai Banking Corporation, which was also home to the Saudi International Bank, the Bankodi Sicilia, the South African gold and platinum mining companies of Johannesburg Consolidated Investment and Barnato Bros, the Taai Bank and the Abu Dhabi Investment Authority. Also damaged were the National Bank of Abu Dhabi, the Banque Indosuez, the

Deutsche Bank and the Long Term Credit Bank of Japan (*Guardian*, 26 April 1993). Bombing a clearly English or British heart of empire was tough business in the City of London of the 1990s. Yet it is very likely that the bombing of these 'unhomely' targets had as much symbolic power as a bombing of the Bank of England, Mansion House or the Royal Exchange. It is precisely the presence of such international businesses that forms the foundation of the contemporary City and ensures that its diminished and domesticated idea of empire is translated into a new global context.

The bombings of the City of London set in train a range of emergency security operations which sought to assure local workers and international investors alike that the City was safe and secure. The first bombing in the City resulted in the implementation of 'Operation Rolling Rock' which increased the presence of uniformed police on the streets of the City and introduced random, short-term, road blocks at various City entry points. The second bombing encouraged the development of a long-term strategic plan to protect the heart of British financial and commercial activity. In the wake of the second bombing one senior City businessman suggested that the 'world's leading financial Capital' should erect a modern version of the medieval London Wall with steel security gates (*The Times*, 27 April 1993). The Corporation of London argued against such heavy-handed security measures but, in conjunction with the City Police, did introduce armed police barricades at all the road entry points to the City ... They called this security plan the 'Ring of Steel'. A second strategy called 'Camerawatch' encouraged private businesses to install their own video surveillance equipment to be linked to a central monitoring service. The aim of 'Camerawatch' was to ensure public areas within the City were monitored '24 hours a day 365 days a year' (City of London Police, 1994, p.2). Installation guidelines were issued by the Corporation to ensure that surveillance cameras were 'visually

discreet' and did not adversely affect 'the character and general ambience of City streets' (Corporation of London, 1993b).

... The City of London's Ring of Steel' and its 'Camerawatch' surveillance system was a strategic response to a residual colonial predicament. It was specifically directed at keeping the IRA out of the City but not so that the City could return to some 'pure' space of the imperial imaginary. Rather, it was a strategy that guarded the City so that it might continue to negotiate its path towards the increasingly cosmopolitan requirements of being a 'global city'. Indeed the 're-walling' of the City of London sought to ensure a secure space for growth that would not, could not, be confined to the traditional geography of the square mile and was already spilling across these 'walls' into surrounding neighbourhoods.

◆ ◆ ◆

Monuments may be formed from artefacts of the past or they may be made anew. In the planning saga of Bank Junction there is a struggle over how to make a monument to a City no longer clearly positioned at the centre; be it of a fading empire or an unpredictable global system. Although the protagonists clearly had different visions for the development site, they argued with a language and logic that were remarkably similar. No party in this struggle challenged the centrality of the Bank Junction to the City, and no party challenged the centrality of the City to the international status of the nation. In the end, the opponents were arguing simply about different ways to monumentalise the grandness of a place whose international status was under transformation and possibly threat. This planning 'battle' can actually be read as a thirty-year public ritual of reiteration and verification of a City of power. In this economic and imaginative transformation the idea of empire lived on and shaped the way the City moved into the future. The City of empire may well have constituted its 'pure' sense of Self from distant imperial relations. The City of the 1980s had to constitute its sense of Self, somewhat precariously, around a heartland that was full of the Others of the new global and regional order.

The redevelopment struggle of Bank Junction shows how the global and the local, the new and the old, the market and the vernacular, interacted and became the means by which the City's status and identity were renegotiated during a period of rapid change. It is tempting to equate 'the new' with less embedded, more superficial global forces of obliteration or, at best, an appropriative mimicry of the authentically local. Equally, it is tempting to align 'the old' with the diminutive, the local, a presumably more authentic embeddedness or possibly even resistance. The struggle of Bank Junction unsettled these alignments. Here local 'resistance' to change resonated with the reactionary nostalgias of royalty and a yearning for the purity of the idea of empire. Here 'change' itself became a nostalgic gesture towards a time when the City did more surely centre its geography around Bank Junction. Here a loyalty to the preservation and enhancement of the local built form actually worked to consolidate the decentring of historical geographies of power and the 'invasion' of the Heartland by a feared Other.

References

Amery, C. and Cruickshank, D. (1975) *The Rape of Britain*, London, Elek.

Anderson, R. (1988) 'Meaning in the urban environment', unpublished PhD, Oxford Polytechnic, Centre for Urban Design.

Baker Harris Saunders. (1988) *No. 1 Poultry Public Inquiry: Proof of Evidence*, London, Department of Environment.

Balfour, A. (1990) *Berlin: The Politics of Order, 1737–1989*, New York, Rizzoli.

Budd, L. and Whimster, S. (eds) (1992) *Global Finance and Urban Living: A Study of Metropolitan Change*, London, Routledge.

Cain, P.J. and Hopkins, A.G. (1993) *British Imperialism: Innovation and Expansion 1688–1914*, London, Longman.

Cassis, Y. (1988) 'Merchant bankers and City aristocracy', *The British Journal of Sociology*, vol.39, pp.114–20.

City of London Police (1994) *Camerawatch: Closed Circuit Television Scheme: Code of Practice*, London, City of London Police.

Clarke, W. (1989) *The City of London Official Guide*, London, Hobson Publishing.

Corporation of London (1986) *City of London Local Plan*, Guildhall, London, Department of Architecture and Planning, Corporation of London.

Corporation of London (1993a) *Schedule of Development*, Guildhall, London, Department of Architecture and Planning, Corporation of London.

Corporation of London (1993b) *Security Camera, Planning Advice Note 1*, Guildhall, London, Department of Planning and Building Security, Corporation of London.

Corporation of London (1993c) 'Business as usual for the city', Press Release, 24 April.

Corporation of London (1994) *The City of London to the Year 2000 and Beyond: Prospects for Office Demand*, Seminar proceedings, 16 May, London, The Royal Institute of Chartered Surveyors, Investment Property Forum and The Corporation of London.

Cullen, G. (1961) *Townscape*, London, The Architectural Press.

Daniels, S. (1989) 'Marxism, culture and the duplicity of landscape' in Peet, R. and Thrift, N. (eds) *New Models in Geography*, vol.II, London, Unwin Hyman, pp.196–220.

Daniels, S. (1993) *Fields of Vision: Landscape Imagery and National Identity in England and the United States*, Cambridge, Polity Press.

Davies, H. (1992) 'Thoughts from a "dangerous man"', *The Spectator*, 5 September, p.12.

de Cronin Hastings, H. (1944) 'Exterior furnishing or sharawaggi: the art of making urban landscape', *The Architectual Review*, January, pp.3–8.

de Cronin Hastings, H. (1945) 'Programme for the City of London', *The Architectural Review*, June, pp.158–87.

DEGW (1985) *Accommodating the Growing City*, Report by DEGW for Rosehaugh Stanhope Plc, London, DEGW.

de Wolfe, I. [de Cronin Hastings, H.] (1949) 'Townscape', *The Architectural Review*, December, pp.355–62.

English Heritage (1988) *Proof of Evidence: No. 1 Poultry Public Inquiry*, London, Department of Environment.

Esher, L. (1983) *A Broken Wave*, London, Pelican Books.

Goodhart, C. and Grant, A. (eds) (1986) *Business of Banking*, London, Gower Press.

Harris, J. and Thane, P. (1984) 'British and European bankers 1880–1914: and "aristocratic bourgeoisie"?' in Thane, P., Crossick, G. and Floud, R. (eds) *The Power of the Past: Essays for Eric Hosbawm*, Cambridge, Cambridge University Press, pp.215–34.

Healey, P. (1990) 'Understanding land and property development processes: some key issues' in Healey, P. and Nabarro, R. (eds) *Land and Property Development in a Changing Context*, London, Gower, pp.1–14.

HRH The Prince of Wales (1989) *A Vision of Britain: A Personal View of Architecture*, London, Doubleday.

Jencks, C. (1988) *The Prince and the Architects and New Wave Monarchy*, London, Academy Editions.

Johnson, C. (1978) *Mies van der Rohe*, New York, Museum of Modern Art.

Jones Lang and Wootton (1994) *Quarterly Review London City Offices, First Quarter*, London, Jones Lang and Wootton

King, A.D. (1990) *Global Cities: Post-Imperialism and the Internationalisation of London*, London and New York, Routledge.

King, A.D.(1991) 'Introduction: spaces of culture, spaces of knowledge' in King, A.D. (ed.) *Culture, Globalization and the World-System: Contemporary Conditions for the Representation of Identity*, Basingstoke, Macmillan in association with the Department of Art and Art History, State University of New York at Binghamton, pp.1–18.

Kutcher, A. (1976) 'The views of St Paul's Cathedral' in Lloyd, D. *et al.* (eds) *Save the City*, London, SPAB.

Lisle-Williams, M. (1984) 'Merchant banking dynasties in the English class structure: ownership, solidarity and kinship in the City of London, 1850–1960', *The British Journal of Sociology*, vol.xxxv, pp.333–62.

London Planning Advisory Committee (1993) *The London Office Market: 1993 Update Report*, London, LPAC in conjunction with Investment Property Databank and Applied Property Research.

McDowell, L. and Court, G. (1994) 'Performing work: bodily representations in merchant banks', *Environment and Planning D, Society and Space*, vol.122, pp.727–50.

Marks, S. (1984) *Corporation of London: Appeals by Number One Poultry Limited and City Acre Property Investment Trust Limited*, Report of the Inspector to the Secretary of State for the Environment, London, Department of Environment.

Matless, D. (1990) 'Ages of English design: preservation, modernism and tales of their history', *Journal of Design History*, vol.3, pp.203–12.

Plender, J. and Wallace, P. (1985) *The Square Mile: A Guide to the New City of London*, London, Century Publishing.

Pryke, M. (1991) 'An international city going "global": spatial change and office provision in the City of London', *Environment and Planning D, Society and Space*, vol.9, pp.197–222.

Pryke, M. (1994) 'Looking back on the space of a boom: (re)developing spatial matrices in the City of London', *Environment and Planning A*, vol.26, no.2, pp.167–332.

Richard Saunders and Partners (1986) *Office Floor Space Survey*, London, Richard Saunders and Partners.

Robins, K. (1991) 'Tradition and translation: national culture in its global context' in Corner, J. and Harvey, S. (eds) *Enterprise and Heritage: Cross-currents in National Culture*, New York and London, Routledge, pp.21–44.

Royal Commission on Local Government in Greater London (1962) *Written Evidence, 1164,* 1959–1960, p.xviii.

Sassen, S. (1991) *The Global City: New York, London, Tokyo,* Princeton, NJ, Princeton University Press.

Stafford, L. (1992) 'London's financial markets: perspectives and prospects' in Budd, L. and Whimster, S. (eds) *Global Finance and Urban Living: A Study of Metropolitan Change*, London, Routledge, pp.31–51.

Thrift, N. (1994) 'On the social and cultural determinants of international financial centres: the case of the City of London' in Corbridge, S., Martin, R. and Thrift, N. (eds) *Money, Power and Space*, Oxford, Blackwell, pp.327–55.

Thrift, N. and Williams, P. (1987) (eds) *Class and Space: The Making of Urban Society*, London, Routledge.

Tugnutt, A. and Robertson, M. (1987) *Making Townscape*, London, Mitchell.

Valuation Office, Inland Revenue (1988–1992) *Property Market Reports*, London, Inland Revenue.

Worskett, R. (1988) *Proof of Evidence: No.1 Poultry Public Inquiry*, Public Inquiry statement, London, Department of Environment.

Wright, P. (1991) *A Journey through Ruins: The Last Days of London,* London, Radius.

Source: Jacobs, 1996, pp.38–69

CHAPTER 4
On space and the city

by Doreen Massey

1 *On the city*

At the end of the last chapter you read an article by M. Ismail Serageldin, entitled 'A decent life' (Extract 3.3). What impression did it make on you? How did it make you feel? Four things in particular remain with me:

1   The simple enormity of the rate and size of urban growth which we are facing.

2   That this is taking place in a context not simply of inequality but of systematically *increasing* inequality: in 1960 the richest 20 per cent of the world's population was thirty times as wealthy as the poorest 20 per cent; it is now *sixty* times as wealthy.

3   That these inequalities extend well beyond the figures for income and wealth and permeate every aspect of how all these billions of people live their lives in the city: 1.3 *billion* people every day live in air deemed unfit to breathe.

4   That there was this sentence: 'A most serious problem is that the rich – whether rich countries or rich individuals – are disengaging themselves, turning their backs on the poor.' It is as though the kinds of disconnection that we have been noting (remember 'structurally irrelevant people'?) in the economic and social sphere are mirrored here in aspects of the political. 'Solutions are feasible,' urges Serageldin. 'The question is whether we will have the political will to act ...'

Sometimes, when I read of such issues at a global scale (and we do now have to think, and imagine, at a global scale), I can feel simply overwhelmed. Are cities of the scale and complexity which we are facing now sustainable either environmentally or socially?

And yet, and yet ... there has also been another side to the city which has permeated what you have read in this book so far. This is the city as a place of excitement, as a place where cultures meet and new things are spawned, the city as a crucible of new ideas and activities, a place where people can escape the sometimes constricting and claustrophobic rules and regulations of a small 'community', the city as a hope for humanity's future. In Chapter 1 you were introduced to the idea of the city as essentially paradoxical; here we have the paradox on its grandest scale. Is 'the city' to be seen as holding out some vision of utopia, or does it inevitably harbour within it elements of our most dystopic fears? Perhaps both things are true.

This short chapter aims to do two things. In section 3 we will explore some of this double-sided nature of the city, some of the tensions which cities embody. In order to do this, however, we need an approach to studying the city, a way of understanding the generation of these tensions and of appreciating their complexity. A first step towards developing such an approach is taken in the next section.

2 *On space and the city*

This book – and the series of which this is the first volume – adopts a theoretical perspective which lays emphasis on the fact that cities, both individually and in the relations between them, are spatial phenomena. On immediate reading, this might seem to be blindingly obvious. Of course cities are 'spatial': they *are* spaces, they exist *in* space. These things are true but we mean something more than this. For one thing, there has in recent years been a sea-change within the social sciences in how space itself is conceptualized. Increasingly the spaces through which we live our lives, and through which the world – and cities – come to be organized are understood as being social products, and social products formed out of the relations which exist between people, agencies, institutions, and so forth. We might see such social relations in terms of the (dis)connections we have been discussing in this and earlier chapters. One way of understanding cities, then, might be as particular patterns of such connections set within wider patterns of the relations with other cities and with the rest of the world. For another thing, we want to argue that by thinking spatially, and through taking account of changes in the way in which space is imagined and organized, we may contribute something specific, and useful, to an understanding of what Serageldin called in his article 'the coming urban century'.

Let us, then, begin by outlining three elements of what can be meant by thinking the city spatially.

1 *The city as specifically spatial*

A number of times in the previous chapters cities have been characterized in terms of their *intensity*. Chapter 1 talked of the way in which cities congregate and combine people and activities, and Mumford coined the term 'geographical plexus' to try to capture this. Chapter 1 (section 1.2) emphasized 'the sheer quantity of possible social interactions' in the city and cited Mumford as arguing that 'the city *like nowhere else* brings people together'. Chapter 1 also began to explore what is meant here by the term 'intensity'. It spoke of two aspects: those of time-and-space (the city as speeded-up interconnections set within an ever spatially-spreading web of external connections) and those of size (see section 2.2). Later it elaborated on the notion of size itself, differentiating between density (section 3.2) and heterogeneity (section 3.3). In our attempt to capture even more of this notion of intensity we might add, say, the simple density of built space, or the city as a captor and transformer of nature ('the city is never distant from nature … it [exists] only by manufacturing itself and nature anew' (Chapter 1 section 2.3)). But perhaps what is most important is that this intensity is something which *emerges as an effect of* all these constellations and intersections. And Chapter 2 drew out how this intensity is also something which is *felt,* in what it termed 'this expressive side to city life'. Intensity is more than sheer size, or density, or heterogeneity alone. As Chapter 1 put it, 'intensity is the result of what happens when large numbers of people are brought together ... it emerges out of the social interactions within the city ...' (section 4).

- We might, then, initially conceptualize the city, as a specifically spatial phenomenon, as a region of particularly dense networks of interaction, from which emerge intense effects, set within areas where interactions are more sparse and spaced out.

We can, moreover, go further than this. Chapter 2 in particular looked inside this intensity and discovered 'many worlds': the multiple stories and rhythms of the city. The shift of rhythms, and of those who live them, as dawn opens out into morning proper, the coexistences of formal and informal worlds in Nairobi and São Paulo, the rush of urgent business past the patience of the beggar. Different social stories, with distinct rhythms, and which create and weave together their own spaces: 'the daily rhythms and movements of cities routinely code and divide city space' (Chapter 2 section 2.1).

- Within the generalized intensity of the city we can detect distinct space–times.
- And these different stories, different space–times, may meet up and affect each other, may repel each other, may overlap in indifference. They coexist in a continuous dance of space–time configurations.

But they are not all equal. That quotation cited a moment ago, when cited fully reads: 'Rather the point is that the daily rhythms and movements of cities routinely code and divide city space on an unequal basis that is rarely acknowledged.' In Anhangabaú the business and professional classes, with the onset of the formal working-day, stride about the streets as if they owned them – as if their's were the only stories to be told. They make it *their* space–time (the-street-of-the-formal-working-day). Through numerical dominance, through street police and private guards, through the very confidence with which they walk, 'they attempt to erase the traces of others'; but the other stories still live on, to emerge in other places, at other times. The Filipina women, their lives submerged all week in domestic labour for others – their lived time–spaces hidden from public recognition – erupt into full view on Sundays, outside the Catholic church (Chapter 2 section 2.3).

Intensities, then, and internally differentiated ones; but, as Chapter 3 went on to stress, cities are also *open* intensities. Each city lives in interconnection with other cities and with non-city areas: 'Chicago was at the hub for many kinds of *connections*: connections which stretched out from the city;

connections which drew people together within the city' (Chapter 1 section 2.2). The complex of internal connections is what gives the potential for intensity; the external connections feed the city with new energies (from food to newly arriving cultures and ideas). Moreover it is through this spatial openness and interconnectedness of cities that they are both held together in interdependence and enabled to pursue their distinctive individual trajectories.

- What we are searching for is a geographical imagination which can look both within and beyond the city and hold the two things in tension.

Such a way of looking at a city – as a kind of *open intensity* – is very abstract. But it does immediately capture some important things. First, it enables us in general terms to imagine the complexity of the many worlds of the city. As Chapter 2 began by stressing we want both to appreciate the multiplicities and the disconnections of the city and to be able to grasp the city as a whole. Second, and to make the concept a little less abstract, it is important to remember that the connections within and between cities are real, hard-won, grounded processes (the Spanish slogging through the mountains to Tenochtitlán, the effort it can cost to establish dialogues across differences, the risky investments of small farmers to build a railway to Chicago); it is through these real interactions, rather than through some rather abstract, single, global system, that cities are both held together and maintain their individualities. And, third, this way of thinking of cities lays a stress on movement, fluidity and 'mixity' in such a way that it becomes apparent that any approach to urban governance and urban planning, say, cannot proceed on the basis of some final, formal plan, nor work with an assumption of a reachable permanent harmony or peace. The order of cities is a dynamic order. What is necessary is a way of approaching this fluidity, openness, and density of interaction: a thinking about process. Such cities are challenges to democracy.

ACTIVITY 4.1   Just pause for a moment and think about that last sentence. In what ways are cities 'challenges to democracy'? Simply pondering our definition of 'cities as specifically spatial' might set you thinking. For instance:

- What kinds of problems, as well as possibilities, might be posed by the city as an intensity?
- Why is it necessary to think about processes rather than about static city plans? ◆

2   Spatial configurations as generative

The fact of cities having a particular spatial form is, however, only the first step in the argument. For what we have seen over and over again in the chapters in this book is that this highly particular spatiality produces effects. Thus, right at the beginning of Chapter 1 we read of the city tending 'to exaggerate ... relationships, if only by bringing them into close proximity' and a little later (section 2.1) we pursued Lewis Mumford's argument that the city is not just a place where lots of things happen to be but, further, that 'the city – by bringing people (and their money) together – both enables new forms of association to be created ..., and also requires of people that they interact in new kinds of ways.' These ideas were developed much further in both Chapter 1 and Chapter 2.

Georg Simmel was one of the earliest modern thinkers to consider these effects. He argued that the effect of concentrated spatial proximity was a necessary social distancing (Chapter 2 section 2.2). Others followed him in this. Thus, in Chapter 1 Louis Wirth's ideas were explored on how the intense spatial proximities of the city might affect people's behaviour. He argued that, ' ... the close physical contact of numerous individuals necessarily produces a shift ... We tend to acquire and develop sensitivity to a world of artefacts and become progressively further removed from the world of nature.' That kind of social distancing of person from person is also described by Sennett in his account of atomistic, enclosed drivers cruising along the freeways of urban USA (Chapter 2, section 2.1).

What is being argued, either implicitly or explicitly, in all these accounts is that *spatial configurations produce effects*. That is, the way in which society (more specifically, the city) is organized spatially can have an impact on how that society/city works. The examples above are really some hypotheses about how the particular spatial form of the city – in particular its intensity – may affect social behaviour and interaction.

Putting the point in this way may help to sort out what could have seemed like a contradiction within the previous chapters. Many times we have insisted upon the multiplicity of the city (indeed I have just done so again) and have argued, for instance in Chapter 3, that it can be problematical to think in terms of a single entity – 'the city' (Chapter 3 section 3.1). The city itself is not, in this sense, a single actor. And yet 'the city' in the specific form of 'cityness' can indeed have effects. It is the latter – an effect of the spatial configuration of the city – which writers such as Simmel, Wirth and Sennett were trying to capture.

Of course, spatial configurations can come in many forms. The writers we have just mentioned were mainly focusing on densities, proximities and juxtapositions. And the impact of such juxtapositions can be detected at levels other than that of the social interaction of individuals. The historic meeting of Aztec and Spanish in the year 1521/One Reed was just such another case: the meeting up, the coming into spatial proximity, of two worlds, two previously separate histories. Not only did this result in the attack on the old city and the emergence of the new one, it also produced Mexico City as a classic mestizo community. These, then, are effects of spatial interconnections and of what we called a couple of times in Chapter 3 'geographical juxtapositions'. As we said there (in section 1.3), 'new "geographical juxtapositions" produce new histories.' So, too, do interconnections over longer distances. The networks of communication, power and influence which connect cities together have their effects on each of them. It is partly interconnections of this sort which make cities the cosmopolitan places they so frequently are.

And *disconnection* can have equally significant repercussions. Castells writes urgently of the inequalities which can be entailed in disconnection, and of the potential social consequences: cities disconnected may struggle to find a new role (both of these were discussed in Chapter 3). Mumford mused gloomily on what might be the effects of the low-density spreading-out of the city (Chapter 1 section 1.2).

● So, we are arguing that cities may be understood spatially and the particular form of the spatial configurations which constitute them will affect 'what happens next'.

ACTIVITY 4.2   Maybe you can pick out other effects of spatiality? Some examples might be: the way it can be used to reinforce difference (Chapters 2 and 3); the notion of 'active place-making events' (J.M. Jacobs); the power of the built environment to hail you as a member or to exclude you as 'out of place' (Chapter 2).

If you were an elected representative of a city or a part of a city, what kinds of issues might you have to cope with which arise from this effect of spatiality? For instance: is it right that certain groups of people are excluded from certain places, or should everyone be able to go everywhere?

This is a huge issue, so don't expect to come to a quick and easy conclusion. Maybe there is no 'general answer'. (It may depend on the places and the groups.) But you might take a couple of examples and think about:

● *how* people are excluded (is it by law, by custom, by hostility, by difficulties of physical access ... ?)

● what are the power relations involved? (Is this the powerful defending their exclusive spaces, or vulnerable groups needing a place to feel safe or at home?) ◆

3 *The openness of the outcome*

It would be mistaken, however, to imagine that a particular spatial form necessarily gives rise to a particular social effect. It is not a simple determinate cause-and-effect relation. We can see that from the chapters we have just read. In contrast to Simmel's pronouncements on the impact of city life, Jane Jacobs described a much more active mixing and interaction. Maybe both things go on. Wirth argued that cities presented the opportunity for people to form new kinds of social interaction – 'bonds that do not rely on kinship ties, neighbourliness, communal sentiments, tradition and "folk" attitudes' (Chapter 1 section 3.1). There is no guarantee that the opportunity will be taken. Richard Sennett's account of the disconnected anomie of the motorized urbanite was a lament, but also something he believed could be changed. And in both Chapter 1 and Chapter 3 it was argued that spatial proximity is not enough to guarantee any particular outcome. For proximity to be turned into a city, something else needs to happen. The article by Ismail Serageldin at the end of Chapter 3 presented a clear case of the openness of such outcomes. The spatial juxtapositions which are 'Third World' (and, indeed, 'First World') cities do not have inevitable outcomes. What can be made of them will depend on resources, on what happens to levels of inequality (in other words, on whether there will be curbs on economic policies which produce inequality), and on political commitment.

The 'answer' to our opening question (are today's cities sustainable either environmentally or socially?), then, is that it depends: it depends on human action. Cities may embody in general terms particular spatial forms, but what is made of them, and what *can* be made of them, and indeed how they can be altered, is up to human actions and ingenuity, and human political will.

3 *On some tensions of urban spatiality*

One way in which spatial configurations are generative in an urban setting is that they precisely produce (though with the reservations mentioned just now) the tensions, or paradoxes, and some of the problems, which we have referred to in previous chapters. It is the sheer density of interactions, and the juxtaposition of so many differences, which generate both the excitements and exhilaration of the city and its potential for cultural innovation, as well as the anxieties and need to withdraw into oneself. The very fact of high population density can 'lead' either to the spread of disease or to the possibility of providing, relatively cheaply, a good local medical service. It will lead to *something*; *what* it leads to is a social choice. It can be the very fluidity and constant movement which spawns a desire for a closed and safe community within it all. The coming-together of differences spatially can generate new mixtures or new divisive hostilities: 'space' can promote contact or be used to divide.

One of the themes that has run through previous chapters is that of community and difference. Cities are huge juxtapositions: they are home to an enormous variety of people, who vary along a wide range of axes of difference – ethnic origin, class, cultural background, sexual preference for instance. People come to cities for a whole variety of reasons, from positive choice to imposed necessity. The element of movement and migration, within and between, is an important aspect of today's city. But cities are also settlements, and a continuing question – a constant process of negotiation – necessarily has to concern how such differences are to live together. How are community and difference to be composed in the internal organization of a city? What is the meaning of 'community' in such a context? Are communities necessarily spatial? And do individual people each only belong to one such community, or to many? Some people come to the city precisely to escape one community (a village perhaps) and to search out another (a gay community maybe).

What, in other words, is the nature of the sociability of the city (and what might it be in a better world)? Is a city to be a collection of spatially separated groups or, at the other end of the spectrum of possibilities, a completely individualized mixture? And while we need to think of the terms in which the sociability of the city could/should be constructed, it is also the case that one of the delights of city life, at least for some, can be the very possibility it holds out for anonymity. At one extreme this may mean to disappear completely into the crowd, at the other it may imply nothing more (but still importantly) than sitting in a café where you know absolutely nobody, and watching the world go by, with scarcely a need to interact. How mix the possibilities of connection and disconnection in the city?

Of course, as you may by now be objecting, the answers to such questions, or even approaches to answers, depend on the *terms* of these connections and disconnections. To be isolated voluntarily for an afternoon or two because you feel like being alone is a different world from the isolation of a beggar cut off from all the structures which connect together to form the dominant networks of sociability. Or the worlds of shanty towns which, though locked into networks of connection of their own, are only peripherally and subordinately tied into the connectivities where lie the power, the money and the influence. Disconnection can mean an exhilarating temporary anonymity or a completely disempowering exclusion. Nor is the situation getting any better. Castells argues strongly that the new connections being made between cities, in this era of an increasingly globalized economy and culture, are themselves precisely responsible for the increasing 'disconnectedness' of significant proportions of the world's poor. To all these questions about the future of cities one must add the rider 'for whom?'.

There are, then, tensions between community and difference, and between sociability and anonymity. The fact of 'cityness' creates both opportunities and problems to be addressed. The juxtapositions of the city can enable the formation of communities of choice (Wirth), can encourage an indifferent mingling, can be used to assert and even magnify existing differences (gated communities), to hide them (the cheerful outer face of the Los Angeles mission) or (the inside of that mission) to protect people from each other. Section 4 of Chapter 2 discussed in detail the range of possibilities.

ACTIVITY 4.3   Re-read section 3.1 of Chapter 1 (it is very short!).

- We stressed above that it is not 'space itself' which influences what happens in the city but the spatial configurations of social relations. There are two reflections on this in this section of Chapter 1. See if you can identify them (they are in the second paragraph and the penultimate paragraph).
- Why are there opportunities for 'new kinds of social interaction' in the city?
- Is the 'superficiality and anonymity' possible in the city a good or a bad thing? Don't just go along with what Wirth thinks – what do *you* think?

Within that exploration of community and difference another 'tension of the city' also emerged: that between movement and settlement. Chapter 2 focused strongly on this, on the city as *both* an arena of mingling rhythms (section 2) and of physical embodiment in the built environment *and* a place to settle in security (section 3). This is again a tension of spatiality.

It has recently been argued that with the turn of the millennium we are experiencing a major shift in the way in which spaces, both within and between cities, are actually organized. In the context of globalization – the increasing stretching of economic, cultural, and financial relations around the world – it has been argued that increasingly we are living, not in 'a space of places', but in 'a space of flows'. It is argued that national boundaries are falling before the onslaught of multinational companies, and that 'local cultures' are increasingly being invaded by fashions and innovations from elsewhere. As was briefly mentioned in Chapter 3 it is undoubtedly the case that such things are happening. Some barriers to movement (flow) are certainly crumbling. Yet we have also seen in this book that there are tendencies in the opposite direction – barriers of many kinds continue to be erected. So it is by no means all 'a space of flows'. As some barriers and boundaries fall, others are built up. It is in part a mirror of this tension between movement and settlement. And it is a specifically spatial tension. It concerns the numerous ways in which we create the spaces and places of our urban world, and the power relations within which we do so. What is at issue is how this mixture of movement and settlement is negotiated and who has power within that process.

ACTIVITY 4.4   Take one of the following and explore:

● the power relations through which these spaces can be constructed

● how the spaces in turn can themselves affect social relations.

1   the shift from mingling in the street to driving along freeways in cars

2   gated communities

3   zoning regulations to separate land uses

4   the formation of 'community areas' – a gay village or a little Sicily. ◆

We could continue for some while detecting and analysing yet more tensions, or paradoxes, of city life. But rather than producing a list, let us conclude by analysing one in detail. The focus for this will be an impassioned piece of journalism – 'Exit from the city of destruction' by Jonathan Glancey. The article is about London, but could be about any city which in recent decades has experienced the impact of neo-liberal economic policies. Indeed it has many resonances with Ismail Serageldin's plea for a decent life which you read in relation to Chapter 3, but which focused primarily on cities of the Third World. 'Neo-liberal' economic thought has been another of those sets of ideas which have travelled around the world attaining a kind of accepted dominance (in the way that it seems like 'common sense') since the mid-1970s. Glancey wishes to raise a voice against it precisely because of its effects (as he interprets them) on cities.

Remember how, in *La Plaza de las Tres Culturas* in Mexico City, the very architectural style of the modern buildings was expressive of a particular economic strategy (industrialization) which itself was emblematic, to the Mexican government and indeed most others, of becoming 'modern' (Chapter 3 section 1.3)? Just so in London, Glancey argues, being part of the up-to-the-minute, anti-state, deregulated world of neo-liberalism is reflected in the physicality of the city – in Day-Glo buses (rather than them all being the red ones we used to know so well), security video cameras, and run-down public infrastructure.

Moreover, although he doesn't put it in those terms, Glancey is also calling for a distinction to be made between a powerful city and a successful city (again, see Chapter 3, this time section 2.2). Glancey is writing of the London of the 1980s and 1990s, a period when it would be hard to deny London's power within a whole range of global hierarchies – finance and banking, culture and the arts, and, indeed, the spreading of the economic gospel of neo-liberalism. Glancey's point is precisely to ask 'for whom?' and to point to those excluded from the beneficial effects of London's global significance – its being 'a global city'. Note that some people are indeed *excluded* – the homeless, the ones whose wages are cut, who lose their jobs, who see former training programmes cut back or abandoned. Just as in the situation described by Castells, the forging of new connections by some can lead to the active disconnection of others from the gains which are expected to accrue. Glancey is acutely aware of these processes of exclusion – indeed, as he reflects, the way we live our lives in the city makes each and every one of us, as we go about our daily lives, 'a bag of contradictions'. What he urges that we need instead is an active concept of 'the public good', of 'civic society'. (You will be asked in a moment what you make of this.)

At his most colourful, Glancey poses the choice between the utopian and dystopian possibilities for the city: the Celestial City and the City of Destruction. Although he agrees we will never reach the former (as was said in section 2, static formal imaginings can never cope with a city which is a dynamic open intensity), at least – he pleads – we should aim for it and avoid the horrors of the City of Destruction. Within this most generalized of alternatives, however, Glancey's argument focuses in around one specific tension: that between order and disorder.

ACTIVITY 4.5   Now study Reading 4A, 'Exit from the city of destruction' by Jonathan Glancey. Then go through it again responding to the following questions.

- What is the distinction, for Glancey, between 'order' and 'disorder'? How, roughly speaking, is he defining the two terms?
- Why does he prefer 'order'?
- Why might we interpret the order/disorder couplet, not so much as a simple choice, but as a tension between the two?
- What is your own position on his arguments? ◆

Glancey's argument is, on the whole, an argument for order as opposed to disorder. By the latter he is primarily referring to the product of the wholesale deregulation and privatization which has occurred since the 1980s. By 'order' he is trying to capture a desire for co-operation, for public sector co-ordination of services and assurance of quality, a commitment to a notion of the city as a whole. He writes of 'the cohesive city' and contrasts 'civic society' with 'an urban miasma of individuals'. His arguments against disorder are based in part on what he sees as the cheapening effect of market competition and in part on the blatant inequality of it all – the inclusions and exclusions which were mentioned earlier. He makes, in my opinion, a very strong case. There are elements of his argument which might be put down to personal preference – is a mixture of bus colours inherently 'worse' than all buses being red, for instance? Yet even here he is trying to get at something deeper – at what that apparently simple difference in colour schemes is all about. The present variety of colours is a physical witness, on the streets, to the glories of – and to the dominance of the ideology of – competition. But, argues Glancey, 'There is no need for a bus to be the colour of Jacob's coat to prove that it belongs to the world of the free.' A really 'free' bus would be part of a good service, available to all everywhere, with decently paid and well-trained drivers, and better maintained than is currently the case.

Thus far, you may be following Glancey and agreeing with him, or you may be quite definitely dissenting. Stay with it, however, and let us pursue the argument a little further.

First, did you agree with his definitions of 'order' and 'disorder'? I did in a way – I knew what he was getting at. But I also harboured a reservation. As in the case of all the tensions in the city, you have to ask 'for whom?'. Here, then, we have to ask 'whose order?'. People tend to view as 'orderly' things either which they themselves control, or which they understand or can easily negotiate. (The Bombay bazaar might look to the uninitiated like a complicated chaos, but – to those who know it – it has its own rhythms and patterns and rules.) Perhaps the companies which have benefited so handsomely from privatization and deregulation see Glancey's downward spiral of a free-for-all-all, as 'a system' which they understand very well – *'the order of the market'.* Likewise, we must ask – and will do so – whose 'order' is embodied in 'the cohesive city' for which Glancey longs?

Second, then, and following directly on, maybe there isn't a simple choice between order and disorder. The question of 'whose order?' already raises this possibility, but there are also other things to consider. Glancey himself recognizes that it is not a simple case of order = good, disorder = bad. The two are often mixed up, more ambiguous, or linked in counterpoint. Tokyo is '*exciting but* physically chaotic' (my emphasis) – the disorder *is* exciting. He recognizes the dangers than can come from (his kind of) 'order' – rigidity, complacency, the potential difficulties of bureaucracy. Re-read now the paragraph beginning 'Disorder can, of course, produce variety ...'. It is a paragraph which encapsulates the ambiguity of it all. He derides the 'passive', 'space-consuming' supermarkets-plus-car-ownership life (exactly as does Sennett) and warms to the 'messy vitality of street markets' (shades of Jane Jacobs). The latter is equated with disorder and yet is loved for its 'variety', 'excitement' and 'hit-and-miss beauty'. Disorder, here, is fun. Ambivalence piled upon ambivalence, then, for one might also question – as we did above – the simple equation anyway of the street market with 'disorder'.

And yet, the issues may be tricky to formulate but Glancey is getting at something real. He recognizes some of the complexity – the long-term plan he craves for would be 'a very gentle one'. Which leads to the crucial question: given that every form of order is 'someone's order', can *any* concept of 'the public good', 'the cohesive city' really be for *everyone?* For, as Glancey puts it, 'every citizen regardless'? Perhaps there are always conflicting interests? In other words, we also have to ask: 'the public good' defined by whom? Glancey's answer lies precisely in that *process of definition* – perhaps what are needed are active democratic processes for defining the public good, recognizing the conflicts over it, and allowing it to change over time. As was argued at the end of the first point in section 2: cities are open intensities … and 'What is necessary is a way of approaching this fluidity, openness and density of interaction: a thinking about process. Such cities are challenges to democracy.' For Glancey the question is: 'Can we create this vision of the democratically ordered but vibrant and diverse city?' And, in the same spirit as Serageldin ('The question is whether we will have the political will to act …'), he answers, 'If we want to, of course we can.'

4 *And finally ...*

One of the strongest themes of this volume has been that cities are the intersections of multiple narratives: the stories which came together in Chicago, the intermingling of rhythms in São Paulo, the long-distance interconnectedness of Madrid and Tenochtitlán. We have also stressed that individual cities have distinctive stories to tell; they have their own trajectories. One of the most significant advantages of 'thinking spatially' is that it enables us to see these different narratives as genuinely co-existing.

This is an important point, and integral to our way of approaching cities in this series. Consider for a moment one of the ways of understanding cities which was outlined, and criticized, in Chapter 3 section 2.3 – the telling of the story as a linear progression from Athens to Los Angeles. (Can you remember what our reservations were?)

One of the evident problems with that approach is that it tells a very 'western' story. We asked in Chapter 3 where other cities – Calcutta, Samarkand – might fit in. Now, one way in which such cities are, frequently and retrospectively, fitted in to such a linear view of urban development is to place them somewhere along the line of progression. They are 'not yet' Los Angeles. We would argue, as indeed we did in Chapter 3, that we would not necessarily *expect* them to be: that they may have their own distinctiveness, their own stories to tell.

There are two reasons to stress this again now, at the end of this book. First, placing cities on a single line of development like that reduces genuinely spatial differences to mere place-in-the-queue. Calcutta on this reading is not so much different as simply 'backward' – in the judgement, of course, of those who drew the line of progression in the first place. In contrast, what 'thinking spatially' enables is the recognition of the contemporaneity of difference. The problems of Calcutta will not be solved, nor its opportunities seized, by some simple process of 'catching up'. The differences between (and also within – remember the structurally irrelevant ones) cities continue to be produced *now*. And that is the second point. We should, in other words, not quash spatial differences by blithely assuming that they are temporal (that is, reduce them to a point on a line of historical progression). That way we avoid facing some of the most important social issues of the new millennium.

REFERENCE

Glancey, J. (1996) 'Exit from the city of destruction', *The Independent*, 23 May, p.20.

READING 4A
Jonathan Glancey:
'Exit from the city of destruction'

If we continue to give in to the politics of selfishness, the modern city will disintegrate into ever-smaller splinters

Today, those of us with money and a degree of health and security are offered an ever increasing choice, not only of things, but of ideas and ways of ordering our lives. The free market enables those who live in cities to satisfy our apparently insatiable and urgent demand for whatever we want – a quarter-pounder with cheese, London buses the colour of a packet of Refreshers, 15 pounds' worth of unprotected sex, sushi and Thai noodles at four in the morning, the occasion to play the good Samaritan dropping the odd coin into the lap of the homeless on London's Hungerford Bridge, and the rich, and ultimately indigestible, recipe of fashionable bars, cafés and shops that those lucky enough to be in work can afford.

Whatever we want, whenever we want it: that's the magic of the modern city. In a free market everyone's choice is valid. There are no longer powerful trade unions or entrenched restrictive practices to hold back entrepreneurs from the pursuit of a quick buck. We can all buy shares and enjoy a bit of the profits ourselves. We are freed from the nuisance of having to think about that elusive and Victorian concept: the public good. There are no rules to tell us what is good or bad. The freedom of choice we seek means, ultimately, that nothing is better, just more expensive.

This notion of infinite choice has been the underpinning of the cities that successive governments have helped to build in Britain over the past 15 years; cities, but not necessarily communities. No city, of course, can be perfect. Christian,

the hero of John Bunyan's *Pilgrim's Progress*, sets out on a journey from the City of Destruction to find the Celestial City that stands behind closely guarded gates on Mount Zion. That city is the City of God and is reachable only after death. This, however, has never put off the living from trying to beautify cities. To do so, from the earliest recorded cities in what today we call Iraq onwards, a degree of planning and co-ordination was necessary. No beautiful city has ever come from the workings of the free market. A perfectly free-market city might be profitable, but never a thing of beauty. It is more likely – Tokyo, for example – to be exciting but physically chaotic. The dream cities we go off to for weekend breaks – Paris, Rome or Siena – to escape our cities of destruction have all been highly planned, or, at least, represent a likeable mixture of regulation and deregulation.

Our cities have often failed to get this mix right. In deregulating and privatising civic services, spaces and utilities, and abolishing the Greater London Council, national government has offered Londoners a diversity of public services and utilities in exchange. Many are cheaper to run and offer lower fares and prices than before.

Many London bus drivers, for example, take home about £150 a week. This means low costs. But is a low-wage urban economy a good thing in the long run? Deregulated bus services may or may not offer a good service. Some do, some don't. What we do know is that drivers no longer receive the expert training they once did under London Transport. The vehicles are shoddy things that get the job done in a perfunctory way. They are no longer the classics of 20th-century design that once made their way into studies of exemplary urban design worldwide.

Deregulation in other areas of the urban economy might also seem to be liberating. By freeing entrepreneurs from minimum wages and maximum hours, for example, the deregulated city offers employment for more people than ever before working in our much celebrated new wave of cafés, bars, restaurants, shops, and clubs. The city can stay open far longer than it did in the days of national and local government hegemony. By keeping wages low, we find jobs for those coming to this country to escape tyranny and poverty abroad. My local car wash was able to cut its price recently, from £8.95 to £5, when it replaced Yugoslavian cleaners with those recently arrived from the Gambia and Nigeria. Should I really be pleased that I have saved £3.95 on washing the car? Whether I should own a car and live in the city centre is another question: we are all a bag of contradictions.

Deregulation and diversity promise choice, but cannot always deliver. Free from restrictive planning and design guidelines, developers and their architects worked up all sorts of fanciful façades and elevations. Competitive tendering, design and build contracts and architectural competitions have all helped to cut the costs once imposed by architects and to shorten design time. Yet the resulting buildings – a plethora of secondhand designs imported from Chicago and New York, have not exactly enhanced the capital.

Meanwhile, what were once public spaces have increasingly been privatised. We live in cities where malls and arcades are heavily policed and locked at night, in which the video camera plays an ever increasing role.

Regulation, order, civic pride and other such concepts might seem old-fashioned. Yet the high-quality public services and utilities, beautifully designed, were never designed solely to delight the eye of the aesthete, architect and connoisseur. They matter because they offer to every citizen regardless of class, creed, colour, age or income the very best we can create and make work at any one time.

Our model should be a well designed civic square, covered market or even a custom-designed red London bus. On its two decks is all of London life, chattering, gossiping, chewing gum, glued to mobile phones. There is no need for a bus to be the colour of Jacob's coat to prove that it belongs to the world of the free. The red London bus, designed and developed over 60 years, offered the highest standards of design and engineering as well as aesthetics for all Londoners and visitors to the capital. It did not discriminate. Design is not some sort of aesthetic bolt-on goodie; it is a way of working for people, of ennobling those we design for. The buses were a part of an integrated and famously well designed public transport network that, from 1933 to the 1980s was co-ordinated and run by London Transport.

That we have agreed to abandon this co-ordinated enterprise is sad, not least because its wilful destruction is a symbol of the way in which we have abandoned the very notion of the cohesive city. The London Transport model as a way of making sense of the modern city may, however, appear to be too forced, too contrived and too limiting. Of course there is a danger that regulating a city may make it too chaste or too rigid. Civic enterprises such as London Transport, the London County Council and its successor, the GLC, were always in danger of becoming complacent, self-regarding bureaucracies run by jobsworths and men and women trained to take no risks.

And yet a city that orders its basic services and utilities, and has a long-term plan, even a very gentle one, for its streets and squares, its parks and river, is a city that is free to breathe freely. And without such a basic order surely a city lacks a spine and the basic components of a nervous system. It cannot hold together: cannot work out when it is in ill-health.

Disorder can, of course, produce variety, excitement and its own hit-and-miss beauty. No dog is more handsome or loyal than the highly deregulated mongrel, while those of us who cannot abide supermarkets and the culture of couch-potato passivity they bring in their space-consuming, car-generating wake, love the messy vitality of street markets.

We support them not only because they offer wide choice and low prices, but because they are part of the civic drama we dream of when we think of sipping an espresso in an open air café in a piazza in Rome or Siena. Why not London, Manchester, Liverpool or Glasgow? A well ordered city provides a beautiful and workable backdrop to the theatre of the street.

This civic drama is an active and not a passive play: cities with a future, as history shows, are highly active transformers creating music and poetry out of chanting and tribal dances, love out of sex, architecture from shelter, art from craft and civic order from rude nature. In Lewis Mumford's words, 'the translation of ideas into common habits and customs, of personal choices and designs into urban structures is one of the prime functions of the city'. A translation the opposite way causes the city to decline and fall.

Order and some degree of regulation do not mean turning London or Manchester into a vision dredged from the note-books of Albert Speer. LCC housing estates from the turn of the century, designed by young socialist architects, still surprise with their gentle and civilised order. Here, were not just so many soulless 'housing units' as we have learnt to call homes for the poor, but a celebration of the ideals of John Ruskin, William Morris and the Arts & Crafts movement: formal, ordered, yet not without beauty, designed to be a decent home to the poorest Londoners, the cockneys of yesterday, the Bengalis of today, and a far cry from either Broadwater Farm or their free-market successors.

Equally, the city with a strong backbone can support the most gorgeous festivities and buildings as wild as Daniel Libeskind's magnificently controversial design for a new extension to the Victoria & Albert Museum. Framed by black cabs, red buses and Giles Gilbert-Scott telephone boxes, Libeskind's buildings will have the power to thrill and yet be kept in its place.

Can we create this vision of the democratically ordered but vibrant and diverse city? If we want to, of course we can. First we have to want a civic society rather than an urban miasma of individuals. And, second, we have to overcome a wish to have as much as we can of everything for as little as possible financially.

If, however, we continue to give in to the politics of selfishness, the modern city will disintegrate into ever smaller splinters, none of them capable of nurturing or providing the big civic gestures, whether Frank Pick's London Transport or the floundering millennium exhibition at Greenwich. These are the gestures that, like giant fire-work displays or music played live in public parks, lift everyday life above the mundane, encourage visitors and long-term business investment and which make us feel we share something in common rather than scurrying around like confused rats in a decaying sewer.

Perhaps, like stubborn children, we have allowed ourselves to be caught in a self-made stew of political dogma and lust for money dressed up as rational economics: if so, we will stay at the bottom of the hill in the City of Destruction with its Day-Glo buses, prostitutes' calling cards, teenagers sleeping rough, public spaces made private, and our only motivation, as passive customers rather than active citizens, a cheaper ride.

This article is an edited version of a lecture on the future of the city given to the Royal Society of Arts.

Source: *The Independent*, 23 May 1996, p.20

Acknowledgements

Grateful acknowledgement is made to the following sources for permission to reproduce material in this book:

Text

pp.69–72: Mai, U. 1997, 'Culture shock and identity crisis in East German cities' in Öncü, S. and Weyland P. (eds) *Space, Culture and Power: New Identities in Globalizing Cities*, pp.76–79, Zed Books; *pp.81–2, 84 and 94–6:* Maspero, F., 1990, *Roissy Express: A Journey Through the Paris Suburbs*, Jones, P. (trans.) 1994, Verso; *pp.90–91:* Caldeira, T. ,1996, 'Fortified enclaves: the new urban segregation', *Public Culture*, vol.8, Duke University Press; *pp.129–30:* Mehrotra, R., 1997, 'One space, two worlds', *Harvard Design Magazine*, Winter/Spring 1997, by permission of Rahul Mehrotra; *pp.130–31:* the text presented here, which focuses on Skid Row in Los Angeles, is extracted from an exhibition at the Getty Research Institute for the History of Art and the Humanities titled 'They saw a very great future here', © 1996, The Getty Research Institute for the History of Art and the Humanities, Santa Monica, CA. Reproduced with permission; *pp.134–6:* Serageldin, I., 1997, 'A decent life', *Harvard Design Magazine*, Winter/Spring 1997, © President and Fellows of Harvard College.

Readings

Reading A: Jacobs, J. M. 1996, *Edge of Empire: Postcolonialism and the City*, pp. 38–69, Routledge, also with permission of the author; *Figure 1:* Neils M. Lund, 'The Heart of the Empire', 1904, © Guildhall Art Gallery, London; *Figure 2:* reproduced from D. Lloyd *et al.*, *Save the City: A Conservation Study of the City of London*, p. 164, © The Society for the Protection of Ancient Buildings; *Figure 3:* No. 1 Poultry © James Stirling Michael Wilford and Associates; *Figures 4, 5 and 6:* reproduced from Jacobs, J. M. 1996, *Edge of Empire: Postcolonialism and the City*, pp. 44–46, Routledge; *Figure 7:* © Roy Worskett; ***Reading B:*** Glancey, J. 1996, 'Exit from the city of destruction', *The Independent*, 23 May 1996, The Independent Newspaper Publishing PLC.

Figures

Figure 1.1: Chicago Architectural Photographing Company/David Phillips; *Figure 1.2:* Topham Picture Library; *Figure 1.3:* reproduced courtesy of Mustoe Merriman Herring Levy; *Figures 1.4 (a) and (b):* Spectrum Colour Library; *Figure 1.4 (c):* Maggie Murray/Format; *Figure 1.4 (d):* Judy Harrison/Format; *Figure 1.5:* Ebenezer Howard, *Tomorrow: A Peaceful Path to Real Reform*, 1898, taken from Peter Hall (1992) *Urban and Regional Planning*, London, Routledge; *Figure 1.6:* Peter Hall (1992) *Urban and Regional Planning*,

London, Routledge; *Figures 1.7 and 1.9:* from *Nature's Metropolis: Chicago and the Great West* by William Cronon, © 1991 by William Cronon. Reprinted by permission of W. W. Norton and Company, Inc.; *Figures 1.8 and 1.10:* Courtesy of Chicago Historical Society; *Figure 1.11:* Collection – Yale University Library; *Figure 1.12:* from *The City* (1925) by R. E. Park, E. W. Burgess and R. D. McKenzie, University of Chicago Press; *Figure 1.14:* S. J. Pile; *Figure 2.1:* © Paul Smith; *Figure 2.2:* Panos Pictures/© Nancy Durrell-McKenna; *Figures 2.3 and 2.5:* Nelson Kon Fotografias; *Figure 2.4:* Panos Pictures; *Figure 2.6:* © Andrew Higgins, photo Richard Jones/Sinopix; *Figures 2.7, 2.9 and 2.11:* Photos by Anaïk Frantz Huppert; *Figure 2.8:* © Editions du Seuil, 1990, in François Maspero, *Les Passagers du Roissy-Express*; *Figure 2.10:* Sue Cunningham Photographic; *Figure 3.1:* © Xavier Miro/Mexicolore; *Figure 3.2:* © Ian Mursell/Mexicolore; *Figure 3.3:* © Sean Sprague/Mexicolore; *Figure 3.4 (a): The Great Temple of the Aztecs* by Eduardo Matos Moctezuma, Thames and Hudson, 1988, Spectrum Colour Library; *Figure 3.5:* Castells, M. 1996, *The Rise of the Network Society, Vol. 1: The Information Age: Economy, Society and Culture*, Blackwell Publishers Ltd; *Figure 3.6:* Biblioteca Apostolica Vaticana.

Table

Table 3.1: Castells, M., 1996, *The Rise of the Network Society, Vol. 1: The Information Age: Economy, Society and Culture*, Blackwell Publishers Ltd.

Cover

Foreground photograph: Reportage Pictures/© Julio Etchart; *background photograph:* Michael Pryke.

Index